Springer Theses

Recognizing Outstanding Ph.D. Research

For further volumes:
http://www.springer.com/series/8790

Aims and Scope

The series "Springer Theses" brings together a selection of the very best Ph.D. theses from around the world and across the physical sciences. Nominated and endorsed by two recognized specialists, each published volume has been selected for its scientific excellence and the high impact of its contents for the pertinent field of research. For greater accessibility to non-specialists, the published versions include an extended introduction, as well as a foreword by the student's supervisor explaining the special relevance of the work for the field. As a whole, the series will provide a valuable resource both for newcomers to the research fields described, and for other scientists seeking detailed background information on special questions. Finally, it provides an accredited documentation of the valuable contributions made by today's younger generation of scientists.

Theses are accepted into the series by invited nomination only and must fulfill all of the following criteria

- They must be written in good English.
- The topic should fall within the confines of Chemistry, Physics, Earth Sciences, Engineering and related interdisciplinary fields such as Materials, Nanoscience, Chemical Engineering, Complex Systems and Biophysics.
- The work reported in the thesis must represent a significant scientific advance.
- If the thesis includes previously published material, permission to reproduce this must be gained from the respective copyright holder.
- They must have been examined and passed during the 12 months prior to nomination.
- Each thesis should include a foreword by the supervisor outlining the significance of its content.
- The theses should have a clearly defined structure including an introduction accessible to scientists not expert in that particular field.

Raquel Alicante

Photoinduced Modifications of the Nonlinear Optical Response in Liquid Crystalline Azopolymers

Doctoral Thesis accepted by
the University of Zaragoza, Spain

Springer

Author
Dr. Raquel Alicante
Departamento de Fisica de la Materia
University of Zaragoza
Zaragoza
Spain

Supervisors
Prof. Dr. Belén Villacampa Naverac
Departamento de Fisica de la Materia
University of Zaragoza
Zaragoza
Spain

Prof. Dr. Rafael Cases Andreu
Departamento de Fisica de la Materia
University of Zaragoza
Zaragoza
Spain

ISSN 2190-5053 ISSN 2190-5061 (electronic)
ISBN 978-3-642-31755-2 ISBN 978-3-642-31756-9 (eBook)
DOI 10.1007/978-3-642-31756-9
Springer Heidelberg New York Dordrecht London

Library of Congress Control Number: 2012945479

Printed on acid-free paper

Springer is part of Springer Science+Business Media (www.springer.com)

To J.

Supervisors' Foreword

The field of nonlinear optics emerged five decades ago, closely related to the invention of the laser. The discovery of new nonlinear optical (NLO) phenomena in inorganic materials generated much interest, owing to their promising role in the development of photonic technologies, and quickly spread to organic materials. A huge number of molecules have been synthesized since the early 1970s, and the microscopic nonlinear parameters have increased by orders of magnitude since then. Along with the progress in the identification of structure–nonlinearity relationships made in the 1990s and 2000s, the practical requirements of transferring high molecular nonlinear response to bulk materials have been one of the main challenges. In particular, the improvement of poling-induced acentric order leading to macroscopic NLO response in polymers together with other properties, such as the reduction of optical losses, have been the focus of a great deal of work. This has borne fruit, and operating electro-optic devices based on guest–host polymers have been designed in recent years. On the other hand, side-chain polymers with attached nonlinear moieties allow a greater concentration of chromophores to be included without any phase separation and provide greater induced order stability than guest–host systems, but their synthesis raises more difficulties. The design of molecules with different functionalities, for instance incorporating photoresponsive units, provided a way to induce polar order at low temperature, via photoassisted or all-optical poling processes, in both guest–host and side-chain systems.

In the thesis work by R. Alicante, side-chain polymers bearing photoaddressable azobenzene moieties and promesogen units are the materials of interest. The work deals with the use of polarized light irradiation and thermal treatments as tools to control the magnitude, the symmetry, and the spatial distribution of the NLO response of photoaddressable polymers. The highly anisotropic nonlinear responses of polymers showing thermotropic liquid-crystalline phases have been interpreted in terms of side-chain intermolecular interactions. The tailoring of the NLO response of guest–host systems using light has been demonstrated even when photoadressability and nonlinearity reside in distinct components. This relevant result allows one to foresee the extension of the advantages of photoassisted poling processes to a wider variety of systems.

Zaragoza, Spain, March 2012

Belén Villacampa Naverac
Rafael Cases Andreu

Preface

Within the framework of electromagnetic radiation interaction, nonlinear optics is a subject covering a series of phenomena in which incident radiation is altered in its frequency, phase, or amplitude while interacting with a medium. Since its birth, linked to the discovery of the laser in 1960, nonlinear optics and photonics have been interrelated, where photonics is understood as the set of technologies that use photons to process information. Typical nonlinear phenomena, such as frequency conversion or optical modulation, have been the key to the development of photonic technology.

Notwithstanding that any material may show nonlinear behavior, there is a list of unquestionable requirements concerning applications. The search for materials showing high and stable nonlinear response, good processability, and, moreover, thermal and photochemical resistance has aroused great interest. At first, the investigation was focused on inorganic materials, which led to results relatively quickly; nonlinear crystals (e.g. KDP, KTP, and BBO) were incorporated in commercial systems and they are often found in frequency doubling and tripling or as parametric oscillators in tunable lasers.

Although organic materials appeared in the 1970s, it was during the 1980s and 1990s that interest in this line of investigation increased hugely. One of the reasons for this was their synthetic versatility, which allows molecules to be designed with specific properties, in particular, with very large nonlinear response. Parallel to the optimization of the nonlinear response, the need to transfer of these large responses efficiently to the macroscopic scale was also highlighted. It seems clear, therefore, that the investigation and development of organic materials for nonlinear optical applications need to use molecular engineering to design and synthesize efficient molecular moieties and materials engineering to achieve macroscopic systems with a molecular arrangement that assures a high nonlinear response.

The difficulty of preparing organic crystals with good quality (moreover, many applications require the absence of centrosymmetry) drew attention to polymeric materials, given their easy processability, fast response and, generally, lower cost. The incorporation of nonlinearity in polymers can be achieved by dispersing nonlinear molecules in them (*guest–host* systems) or by attaching them to the polymer chain chemically, often as side groups. In any case, the mentioned noncentrosymmetric arrangement requires the polymers to be able to be ordered polarly, usually by an electric field that orients molecular dipoles. Unfortunately, many of the most efficient nonlinear molecules have large dipole moment values, promoting molecular interactions that give rise to centrosymmetric structures and thus yielding a nonlinear response poorer than expected. In recent years, one of the objectives in the design of polymeric guest–host systems for electro-optic applications has been to decrease these interactions, while maintaining a high content of chromophore and to look for structures that contribute to stabilizing polar orientation.

Side-chain polymers, in turn, allow a higher chromophore content to be included while avoiding phase separation and they offer better stability in the induced order, particularly when high glass transition temperature (T_g) polymers are used if the orientation process takes place at very high temperatures. As an alternative way to force chromophore orientation, promoting the stability of the established order, the use of liquid-crystalline polymers has been proposed. In these kinds of systems, very anisotropic nonlinear responses have been reported, but the effect of chromophore interactions upon the global nonlinear response magnitude has not been so widely studied. Part of the work presented here falls within this line. The nonlinear response of a number of homo- and copolymethacrylates showing nematic and smectic phases has been studied. Nonlinear chromophores in the side chain, besides having promesogenic nature, include photo-orientable azo groups, which has enabled their orientation to be modified optically and its influence on nonlinear response to be studied. On the other hand, they offer the possibility of using photoassisted processes to orient polarly with an electric field, allowing breaking of centrosymmetry at temperatures well below T_g, owing to the increase in the mobility induced by light. Furthermore, the applied selective illumination in this process leads to nonlinear gratings, which combine an efficient second harmonic generation in thin film configuration with a spatial filtering of 2ω light.

In Chap. 1 of this report, some theoretical concepts and material characteristics (chromophores and liquid-crystalline polymers) on which this thesis is centred are reviewed.

In Chap. 2, devoted to the experimental details, the techniques used to characterize linear and nonlinear optical properties of molecules and thin polymeric films, as well as the experimental data analysis methods, are thoroughly described.

The second-order nonlinear response results, obtained from EFISH (electric field-induced second harmonic) technique measurements for several families of compounds are gathered in Chap. 3, as well as a discussion of these results.

The two following chapters are focused on the nonlinear properties of polymeric films oriented by an electric field. Chapter 4 describes a series of copolymers with different nonlinear and photo-orientable chromophore content in the side chain. The nonlinear response, which can be controlled by thermal and optical treatments, is analyzed by means of second harmonic generation measurements, as a function of the chromophore content. In Chap. 5, in turn, the nonlinear response of highly efficient chromophores mixed with photo-orientable polymers with a slight response is described, proving that polar photoassisted orientation techniques can be extended to these kinds of systems. Moreover, the coupling between the induced order in the polymeric matrix (by light irradiation or heating) and the order achieved by the dispersed chromophores is investigated.

Finally, in Chap. 6 diffraction grating formation is described in azopolymers with donor-π-acceptor chromophores. Furthermore, NLO properties were induced in the mentioned gratings and second harmonic diffraction efficiency was characterized as a function of different parameters, as the amplitude of the surface relief or the different NLO response characteristic of each polymer.

the induced order in the polymeric matrix (by light irradiation or heating) and the order achieved by the dispersed chromophores is investigated.

Finally, in Chap. 6 diffraction grating formation is described in azopolymers with donor–acceptor chromophores. Furthermore, NLO properties were induced in the mentioned gratings and second harmonic diffraction efficiency was characterized as a function of different parameters, as the amplitude of the surface relief or the different NLO response characteristic of each polymer.

Acknowledgments

As a result of the intensity of the years spent working for this thesis, personal and professional elements have often become mixed, and many people deserve to be mentioned here.

First, I want to put on record how lucky I have been to have these two supervisors, Belén Villacampa and Rafa Cases, at my disposal as they love their job, and have reminded me how good it feels to research for the pleasure of knowing. Thanks for your kindness and patience. I also extend my thanks to the "Polymers for Optical Applications" group, with whom I have shared the laboratory from time to time—Rafael Alcalá, Carlos Sánchez, Miguel Lomba, and Cristina Berges,—for creating an excellent working environment.

I also have to thank two groups in the Organic Chemistry Department of the University of Zaragoza. First, the Liquid Crystal and Polymer Group, especially Prof. Luis Oriol, Patricia Forcén, and Eva Blasco, for the synthesis of the materials and for being available every time I needed them, and second, the group of Prof. Javier Garín, for the prolific and efficient synthesis of NLO chromophores, and for their patience in waiting for the results. I also thank them for their sense of humor when it came to discussing the results, which made work much funnier.

On a more personal level, I thank my thesis fellows (who are also my friends), Enrique, Sol, Pepa, Clara, Vera, Sara, and María, for welcoming me as one more "physicist" and for all the times we laughed together. I am sure you will do great and we will meet again. Finally, I want to make a special dedication to Juancho: "Just Breathe", and to my friends María and Lidia as well as to all my family.

Contents

Acronyms and Abbreviations

AIBN	Azobis(isobutyronitrile)
BCOG	Binary Chromophore Organic Glass
BLA	Bond Length Alternation
BP	Bandpass
cc	Complex Conjugate
cgs	Centimetre–gram–seconds (units)
CL	Cyanine Limit
CPHF	Coupled Perturbed Hartree–Fock
CT	Charge Transfer
DANS	4-Dimethylamino-4′-nitro-stilbene
DMF	Dimethyl formamide
DMSO	Dimethyl sulfoxide
DR1	Disperse Red 1
EFISH	Electric Field-Induced Second Harmonic
esu	Electrostatic units
FWHM	Full width at half maximum
GPC	Gel Permeation Chromatography
HH	Herman and Hayden
ICMA	Instituto de Ciencia de Materiales de Aragón
IR	Infrared
ITO	Indium Tin Oxide
JK	Jerphagnon and Kurtz
LC	Liquid Crystalline
LCP	Liquid Crystalline Polymer
LHS	Left-hand side
NIR	Near Infrared
NLO	Nonlinear optical
NMR	Nuclear Magnetic Resonance
NPP	*N*-(4-nitrophenyl)-L-prolinol
OD	Optical density
PAP	Photoassisted poling
PI	Polydispersity Index

PMMA	Poly(methylmethacrylate)
RHS	Right-hand side
RT	Room Temperature
SCLC	Side-chain Liquid Crystal
SHG	Second Harmonic Generation
TCF	Tricyanofuran
TDDFT	Time-dependent density functional theory
TE	Transverse Electric
TM	Transverse Magnetic
UV	Ultraviolet

Chapter 1
Introduction and Basic Theory

In this chapter some basic concepts concerning nonlinear optics (NLO) will be covered as well as theoretical fundamentals on which the central measurements of this thesis are based. The development of the model used to characterize the studied polymeric systems will also be shown in the following pages, in order to make clear the subsequent treatment of the experimentally obtained data.

1.1 Nonlinear Optics

1.1.1 Introduction to Nonlinear Optical Phenomena

In general terms, nonlinear optical phenomena are thought to be those processes in which an interaction between electromagnetic waves occurs (in the optical frequency range) as they propagate through a material medium, giving rise to changes in their frequency, phase or amplitude with respect to the incident ones. Every material can be considered as nonlinear since if it is irradiated with intense-enough light, it will be able to produce nonlinear optical processes. The "ordinary" nonlinear optical response would be an approximation that considers that the polarization P of the material induced by the electric field E is linear with the field:

$$\vec{P} = \varepsilon_0 \tilde{\chi}\ \vec{E} \tag{1.1}$$

where ε_0 is the vacuum dielectric permittivity and χ the susceptibility.

For small values of E the proportionality of the polarization with the field is maintained, as it is necessary to achieve values of E of the same magnitude as the

R. Alicante, *Photoinduced Modifications of the Nonlinear Optical Response in Liquid Crystalline Azopolymers*, Springer Theses, DOI: 10.1007/978-3-642-31756-9_1,

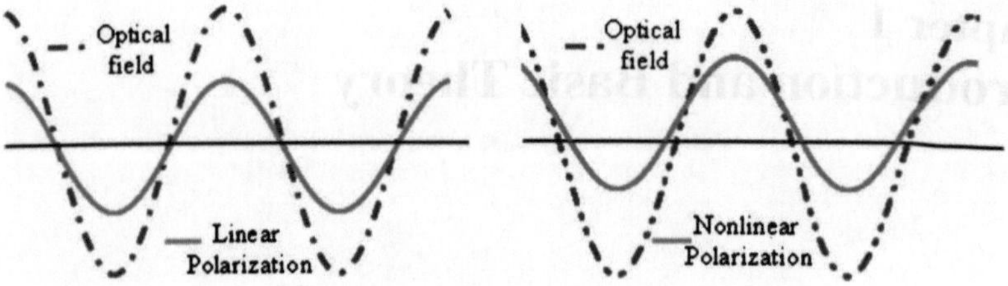

Fig. 1.1 Illustration of the induced linear and nonlinear response vs. field

inner field of matter[1] before superior-order terms in the expression become important [1]:

$$\vec{P} = \varepsilon_0 \left(\tilde{\chi}^{(1)} \vec{E} + \tilde{\chi}^{(2)} \vec{E}^2 + \tilde{\chi}^{(3)} \vec{E}^3 + \ldots \right) \tag{1.2}$$

These field values were inaccessible until the advent of laser technology, so the birth of nonlinear optics is associated with this discovery. When a material is irradiated with very intense light, such as that generated by a laser, linear conditions are broken and new optical phenomena are observed related to the superior-order terms in the previous expression *P*(*E*) (Fig. 1.1).

1.1.2 Microscopic Origin of the Nonlinear Susceptibility

As an approach to the microscopic origin of the nonlinear susceptibility, one usually starts out from the analysis of the linear case according to the classical Lorentz harmonic oscillator model [2]. The medium is considered as a group of atoms, for simplicity each with just a single electron. An electric field induces the separation of positive and negative charges of the nucleus and the electron, and if it changes in a harmonic way with time, the medium behaves like a set of oscillating dipoles [3]. In the linear approach, these dipoles will radiate waves at the same frequency as the field that excites them. Neglecting nucleus movement, the equation of motion for each of the oscillators will be

$$\frac{d^2x}{dt^2} + \gamma \frac{dx}{dt} + \omega_0^2 x = \frac{-e}{m} E, \tag{1.3}$$

[1] $E = \frac{k\,q}{(r_B)^2} = 10^{11} \quad V/m$

$q = 1.6 \times 10^{-19} C$ (electron charge)

$k = 9 \times 10^9\ \mathrm{Nm^2/C^2}$ (proportionality constant)

$r_B = 0.5\,\mathring{A} = 5 \times 10^{-11} m$ (stationary orbit radius)

where x is the displacement of the electron according to the equilibrium position and E is the field acting on the dipole.

$$E = E_0 \cos \omega t = \frac{1}{2}\left(E_0\, e^{i\omega t} + cc\right), \quad \text{where cc = complex conjugate} \tag{1.4}$$

The solution of the previous equation of motion is

$$x = \frac{-eE_0}{2m} \frac{e^{-i\omega t}}{\omega_0^2 - \omega^2 - i\gamma\omega} + cc \tag{1.5}$$

When the external field frequency gets closer to the oscillator's characteristic one, the displacement x increases and it is maximum when $\omega = \omega_0$, which corresponds to a resonance situation.

Bearing in mind that the electric dipole moment is the product of the charge e and the distance x between the charges, and that the macroscopic polarization, P, is the dipole moment per volume unit, its expression will be

$$P = \frac{Ne^2E_0}{2m}\left[\frac{e^{-i\omega t}}{\omega_0^2 - \omega^2 - i\gamma\omega} + cc\right] = \frac{1}{2}\left[\varepsilon_0 \chi E_0 e^{-i\omega t} + cc\right] \tag{1.6}$$

where N is the number of atoms per volume unit. If we compare this expression with 1.1, it can be seen that χ, the linear susceptibility, can be expressed as

$$\begin{aligned}\chi^{(1)}(\omega) &= \frac{Ne^2}{m\varepsilon_0} \frac{1}{\omega_0^2 - \omega^2 - i\gamma\omega} \\ &= \frac{Ne^2}{m\varepsilon_0}\left[\frac{\omega_0^2 - \omega^2}{\left(\omega_0^2 - \omega^2\right)^2 - \gamma^2\omega^2} + \frac{i\gamma\omega}{\left(\omega_0^2 - \omega^2\right)^2 - \gamma^2\omega^2}\right] \\ &= \chi^{real} + i\chi^{imag}\end{aligned} \tag{1.7}$$

It can be seen that the susceptibility is essentially real when the damping, γ, is low (few losses). The real part determines the response of the material in phase with the external field and corresponds to the light dispersion that occurs when light passes through any material medium. The imaginary part, which corresponds to the polarization of the medium in quadrature with the external field, is related to the losses and it is associated with the optical absorption.

When the external fields are very intense, the charges are further away from equilibrium, so the restoring forces that appear are not linear any more with the displacement. If a nonlinear term is added to the motion Eq. 1.3, we obtain the equation of an anharmonic oscillator:

$$\frac{d^2x}{dt^2} + \gamma\frac{dx}{dt} + \omega_0^2 x + ax^2 = \frac{-e}{m}E \tag{1.8}$$

To obtain the solution for x, it is accepted that it can be written as a series expansion [2]:

$$x = x_1 + x_2 + \ldots = a_1 E + a_2 E^2 + \ldots,$$

Substituting for x in (1.8), the terms in the same power of E can be grouped. If we take the linear ones, we will find the linear case, which has been already described, whereas taking the quadratic ones results in the expression

$$\frac{d^2x_2}{dt^2} + \gamma\frac{dx_2}{dt} + \omega_0^2 x_2 + a x_1^2 = 0, \tag{1.9}$$

which can be solved by replacing $x_2 = a_2 E^2$. The solution is the sum of two terms, one oscillating at 2ω frequency (second harmonic), and the other independent of time [4].

$$x_2 = \frac{-ae^2}{4m^2}\left[\left[\frac{E_0^2 e^{-i2\omega t}}{\left((\omega_0^2-\omega^2)-i\omega\gamma\right)^2\left((\omega_0^2-4\omega^2)-2i\omega\gamma\right)} + cc\right] + \left[\frac{2E_0^2}{\left((\omega_0^2-\omega^2)-i\omega\gamma\right)\left((\omega_0^2-\omega^2)-i\omega\gamma\right)\omega_0^2}\right]\right] \tag{1.10}$$

If we use analogous definitions to those described for the dipole moment and nonlinear polarizations, we obtain that the second-order susceptibility can be depicted as a function of the linear susceptibilities corresponding to the different frequencies involved in the process [2]:

$$\chi^{(2)}(2\omega = \omega + \omega) = \frac{ma\varepsilon_0^2}{Ne^3}\chi^{(1)}(2\omega)\chi^{(1)}(\omega)\chi^{(1)}(\omega) \tag{1.11}$$

One general characteristic of any susceptibility is its great enhancement near resonance conditions. Resonance will occur if any of the incident frequencies or those resulting from interaction is close to the system's characteristic one, ω_0.

1.1.3 Nonlinear Polarization

As a result of what has been described in the previous section, it can be understood that when a nonlinear material is irradiated with very intense light, oscillations for every combination of incident frequencies occur. To a first approximation, if we consider an isotropic medium the expression 1.2 can be expressed simply as

$$P(E) = \varepsilon_0\left[\chi^{(1)}E + \chi^{(2)}E^2 + \chi^{(3)}E^3 + \cdots\right], \tag{1.12}$$

where (real) susceptibilities of order m, $x^{(m)}$, are scalar magnitudes. We will develop Equation 1.12 for a monochromatic and linearly polarized incident wave $E = E_0 \cos\omega t$. Then, the induced polarization (up to third order) can be written as

$$\begin{aligned} P(t) &= \varepsilon_0\chi^{(1)}E_0\cos\omega t + \varepsilon_0\chi^{(2)}E_0^2\cos^2\omega t + \varepsilon_0\chi^{(3)}E_0^3\cos^3\omega t \\ &= \varepsilon_0\chi^{(1)}E_0\cos\omega t + \frac{\varepsilon_0\chi^{(2)}}{2}E_0^2 + \frac{\varepsilon_0\chi^{(2)}}{2}E_0^2\cos 2\omega t \\ &\quad + \frac{3\varepsilon_0\chi^{(3)}}{4}E_0^3\cos\omega t + \frac{\varepsilon_0\chi^{(3)}}{4}E_0^3\cos 3\omega t \end{aligned} \tag{1.13}$$

In (1.13),[2] the terms related to the different nonlinear phenomena can be identified. For instance, second harmonic generation corresponds to 2ω oscillation, where it can be seen that its intensity would be proportional to the square of the incident wave. The contributions to $P(t)$ oscillating at different frequencies 0, ω, 2ω and 3ω, can therefore be separated and they are written as

$$P^{(0)} = \frac{\varepsilon_0}{2}\chi^{(2)}E_0^2 \quad (1.14)$$

$$P^{(\omega)}(t) = \varepsilon_0\left[\chi^{(1)}E_0 + \frac{3}{4}\chi^{(3)}E_0^3\right]\cos\omega t \quad (1.15)$$

$$P^{(2\omega)}(t) = \frac{\varepsilon_0}{2}\chi^{(2)}E_0^2\cos 2\omega t \quad (1.16)$$

$$P^{(3\omega)}(t) = \frac{\varepsilon_0}{4}\chi^{(3)}E_0^3\cos 3\omega t \quad (1.17)$$

The term $P^{(0)}$ represents an stationary contribution involving the appearance of a constant potential, which is proportional to the second-order susceptibility. It is an optical rectification process, in which the material responds with a stationary perturbation to the excitation of the oscillating field at frequency ω. The terms $P^{(2\omega)}$ and $P^{(3\omega)}$ correspond to the second and third harmonic generation of the incident light, respectively. The intensity of the two harmonics is proportional to the square and cube, respectively, of the incident intensity. The term $P^{(\omega)}$ includes the linear response $P_L^{(\omega)} = \varepsilon_0\chi^{(1)}E_0\cos\omega t$, together with the nonlinear term P_{NL}^{ω}, which modifies the response at the excitation frequency, causing changes in the refractive index [1].

From here on, we will focus this development on second harmonic generation, since the studies in this thesis are based on measurements of this property.

1.1.4 Second Harmonic Generation

Second harmonic generation (SHG) consists in obtaining light at frequency 2ω from light at frequency ω that strikes a nonlinear material (Fig. 1.2), [3].

SHG was observed for the first time by Franken et al. [5], at the University of Michigan. They focused a 3 kW pulse from a ruby laser emitting at 694.3 nm in a quartz crystal, and they detected light at 347.15 nm. This experiment took place 1 year after the discovery of the laser, and it is considered to be the birth of nonlinear optics, as it was the first experiment to show the possibility of energy exchange between coherent beams.

[2] As $\cos^2\omega t = \frac{1}{2} + \frac{1}{2}\cos 2\omega t$ and $\cos^3\omega t = \frac{3\cos\theta + \cos 3\theta}{4}$

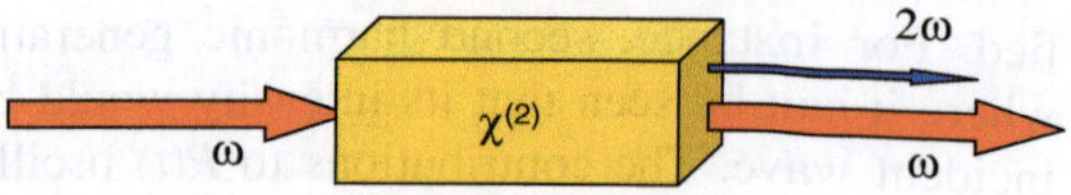

Fig. 1.2 Schematic representation of second harmonic generation

1.1.4.1 Nonlinear Susceptibility Tensor in Anisotropic Media

For an anisotropic medium, the *m*th-order susceptibilities ($\chi^{(m)}$) are $(m+1)$-rank ($m = 1$ linear, $m = 2, 3, \ldots$ nonlinear) tensors and the components of the polarization can be expressed as a function of the corresponding tensor elements:

$$\begin{aligned} P_i^L &= \varepsilon_0 \chi_{ij}^{(1)} E_j \quad (m = 1,\ \text{linear}), \\ P_i^{NL} &= P_i^{(2\omega)} = (1/2)\chi_{ijk}^{(2)} E_j E_k \quad (m = 2,\ \text{quadratic case})\,. \end{aligned} \tag{1.18}$$

In order to simplify the writing regarding notation, we use convention II (see Appendix C) although the experimental values will be expressed using convention I, which is more common in the literature. Then, each one of the subscripts *j*, *k*, covers the three components of the electric field (*x*, *y*, *z* or 1, 2, 3). $\chi_{ijk}^{(2)}$ is then a 3rd-rank tensor (27 components), which turns the action of the two electric vector fields into one polarization vector [6]:

$$\begin{pmatrix} \chi_{xxx}^{(2)} & \chi_{xyy}^{(2)} & \chi_{xzz}^{(2)} & \chi_{xyz}^{(2)} & \chi_{xzy}^{(2)} & \chi_{xzx}^{(2)} & \chi_{xxz}^{(2)} & \chi_{xxy}^{(2)} & \chi_{xyx}^{(2)} \\ \chi_{yxx}^{(2)} & \chi_{yyy}^{(2)} & \chi_{yzz}^{(2)} & \chi_{yyz}^{(2)} & \chi_{yzy}^{(2)} & \chi_{yzx}^{(2)} & \chi_{yxz}^{(2)} & \chi_{yxy}^{(2)} & \chi_{yyx}^{(2)} \\ \chi_{zxx}^{(2)} & \chi_{zyy}^{(2)} & \chi_{zzz}^{(2)} & \chi_{zyz}^{(2)} & \chi_{zzy}^{(2)} & \chi_{zzx}^{(2)} & \chi_{zxz}^{(2)} & \chi_{zxy}^{(2)} & \chi_{zyx}^{(2)} \end{pmatrix}. \tag{1.19}$$

This tensor is symmetric in the permutation of *j* and *k* as the component's order of the fields is irrelevant. By applying this previously quoted symmetry condition with respect to index permutation, $\chi_{ijk} = \chi_{ikj}$, it is fulfilled tha

$$\chi_{xyz} = \chi_{xzy} \quad \chi_{yxz} = \chi_{yzx} \quad \chi_{zxy} = \chi_{zyx}, \text{ etc.},$$

which leaves the tensor reduced to the following 18 components [4]:

$$\begin{pmatrix} p_x^{2\omega} \\ p_y^{2\omega} \\ p_z^{2\omega} \end{pmatrix} = \frac{1}{2} \begin{pmatrix} \chi_{xxx}^{(2)} & \chi_{xyy}^{(2)} & \chi_{xzz}^{(2)} & \chi_{xyz}^{(2)} & \chi_{xxz}^{(2)} & \chi_{xxy}^{(2)} \\ \chi_{yxx}^{(2)} & \chi_{yyy}^{(2)} & \chi_{yzz}^{(2)} & \chi_{yyz}^{(2)} & \chi_{yxz}^{(2)} & \chi_{yxy}^{(2)} \\ \chi_{zxx}^{(2)} & \chi_{zyy}^{(2)} & \chi_{zzz}^{(2)} & \chi_{zyz}^{(2)} & \chi_{zxz}^{(2)} & \chi_{zxy}^{(2)} \end{pmatrix} \begin{pmatrix} E_x^2 \\ E_y^2 \\ E_z^2 \\ 2E_yE_z \\ 2E_xE_z \\ 2E_xE_y \end{pmatrix} \tag{1.20}$$

In the contracted expressions of the polarization, the frequencies are usually included in the argument of the susceptibility tensor, such that the first one is the resulting polarization frequency:

$$P_i^{(2\omega)} = (1/2)\varepsilon_0 \chi_{ijk}^{(2)}(-2\omega; \omega, \omega) E_j(\omega) E_k(\omega) \tag{1.21}$$

Traditionally, most of the experimental information about the second-order nonlinear properties has been obtained from SHG measurements, in which the d coefficient is often used, normally defined as $\chi^{(2)}_{ijk} = 2d_{ijk}$. It is also common to use the contracted notation (as in the piezoelectric tensor), identifying each element with two indices [2] instead of three, so that $d_{ijk} = d_{IL}$, where

I: $1 = x, 2 = y, 3 = z$

and

L	1	2	3	4	5	6
jk	xx	yy	zz	yz zy	zx xz	xy yx

Therefore

$$\begin{array}{llll} d_{xxx} = d_{111} = d_{11} & d_{xyz} & d_{123} = d_{132} = d_{14} \\ d_{xyy} = d_{122} = d_{12} & d_{xzx} & d_{131} = d_{113} = d_{15} \\ d_{xzz} = d_{133} = d_{13} & d_{xxy} & d_{112} = d_{121} = d_{16} \end{array}$$

This would give

$$\begin{bmatrix} P_x \\ P_y \\ P_z \end{bmatrix} = \begin{bmatrix} d_{11} & d_{12} & d_{13} & d_{14} & d_{15} & d_{16} \\ d_{21} & d_{22} & d_{23} & d_{24} & d_{25} & d_{26} \\ d_{31} & d_{32} & d_{33} & d_{34} & d_{35} & d_{36} \end{bmatrix} \begin{bmatrix} E_x^2 \\ E_y^2 \\ E_z^2 \\ 2E_yE_z \\ 2E_xE_z \\ 2E_xE_y \end{bmatrix} \quad (1.22)$$

It is possible to make another simplification in the tensor form in those situations where the Kleinmann condition of symmetry holds [7]. This is based on the fact that when the involved frequencies are far away from the resonance condition of the material, and the energy is exchanged without dissipation in the medium, the value of the coefficients is the same for any permutation of the three Cartesian coordinates.

$$\chi_{ijk} = \chi_{kij} = \chi_{jki} = \chi_{jik} = \chi_{kji} = \chi_{ikj},$$

that is,

$$\begin{array}{lll} d_{14} = d_{25} = d_{36} & d_{12} = d_{26} & d_{23} = d_{34} \\ d_{13} = d_{35} & d_{16} = d_{21} & d_{24} = d_{32} \\ d_{31} = d_{15} & & \end{array}$$

This way, the tensor is reduced to 10 independent components [2]:

$$\mathrm{d}_{IL} = \begin{pmatrix} \mathrm{d}_{11} & \mathrm{d}_{12} & \mathrm{d}_{13} & \mathrm{d}_{14} & \mathrm{d}_{31} & \mathrm{d}_{21} \\ \mathrm{d}_{21} & \mathrm{d}_{22} & \mathrm{d}_{23} & \mathrm{d}_{24} & \mathrm{d}_{14} & \mathrm{d}_{12} \\ \mathrm{d}_{31} & \mathrm{d}_{24} & \mathrm{d}_{33} & \mathrm{d}_{23} & \mathrm{d}_{13} & \mathrm{d}_{14} \end{pmatrix} \quad (1.23)$$

Apart from the already mentioned symmetry conditions, the susceptibility tensor has to describe the structural properties in the nonlinear medium. This results in certain restrictions on the number of nonzero and independent components of the tensor, depending on the crystalline symmetry of the medium. Each crystalline solid belongs to one of the 32 point groups gathered in seven crystalline systems. If we apply one of the symmetry operations characteristic of the crystal to its susceptibility tensor, this must also remain invariant.

For instance, we can consider the inversion operation ($\vec{r} \rightarrow -\vec{r}$,), which transforms the electric field and the polarization as $E \rightarrow -E$ and $P \rightarrow -P$. In a centrosymmetric medium, this operation has to leave the sign and the magnitude of each physical property unchanged, in particular: $\chi^{(2)} \rightarrow \chi^{(2)}$. As a consequence, $P^{(2)} = \chi^{(2)}EE$ is transformed into $-P^{(2)} = \chi^{(2)}(-E)(-E)$, so $\chi^{(2)}$ has to be zero [6]. That is, the $\chi^{(2)}$ tensor is null for any centrosymmetric medium and therefore also in isotropic materials such as gases, liquids or amorphous solids. For each non-centrosymmetric crystal, the tensor in expression 1.22 will have a particular form. As an example, we will consider the case of symmetry class 2 crystals of the monoclinic system, which would correspond to a sample with a binary axis of symmetry.

If the z-axis coincides with the binary axis, when the symmetry operation is applied, that is, for a rotation through 180°, the tensor and the sample will remain unchanged:

$$x \rightarrow -x$$

$$y \rightarrow -y$$

$$z \rightarrow z$$

We will see that the coefficients d_{ijk} with none of the indices corresponding to z, or in which two are z, are going to cancel.

Let us suppose first that the field has only an x component:

$$\vec{E} = E_x\hat{i}, \; E_z = E_y = 0 \; \vec{E} = (E_x, \; 0, \; 0)$$

$$\begin{aligned} P_x &= d_{11}E_x^2 \\ P_y &= d_{21}E_x^2 \\ P_z &= d_{31}E_x^2 \end{aligned} \quad \xrightarrow[2]{\text{Applying180° rotation:}} \quad \begin{aligned} -P_x &= d_{11}(-E_x)^2 \\ -P_y &= d_{21}(-E_x)^2 \\ P_z &= d_{31}(-E_x)^2 \end{aligned}$$

For the previous condition to be fulfilled, $d_{11} = d_{21} = 0$ and $d_{31} \neq 0$.

If we use the same reasoning for the components y and z, so that $\vec{E} = (0, \; E_y, \; 0)$ and $\vec{E} = (0, \; 0, \; E_z)$, we obtain

$$d_{12} = d_{22} = 0 \text{ and } d_{32} \neq 0$$

and

$$d_{13} = d_{23} = 0 \text{ and } d_{33} \neq 0.$$

By assuming that $\vec{E} \equiv (E_x, E_y, 0)$ and by taking into account the nonzero coefficients we obtain that

$$\begin{aligned} P_x &= 2d_{16}E_xE_y \\ P_y &= 2d_{26}E_xE_y \\ P_z &= d_{31}E_x^2 + d_{32}E_y^2 + 2d_{36}E_xE_y \end{aligned} \quad \xrightarrow[2]{} \quad \begin{aligned} -P_x &= 2d_{16}(-E_x)(-E_y) \\ -P_y &= 2d_{26}(-E_x)(-E_y) \\ P_z &= d_{31}(-E_x)^2 + d_{32}(-E_y)^2 \\ &\quad + 2d_{36}(-E_x)(-E_y) \end{aligned}$$

From here, it is deduced that

$$d_{16} = d_{26} = 0 \text{ and } d_{36} \neq 0$$

Analogously, if $\vec{E} = (E_x, \ 0, \ E_z)$ then

$$d_{35} = 0, d_{15} \neq 0 \text{ and } d_{25} \neq 0$$

And, on the other hand, if $\vec{E} = (0, \ E_y, \ E_z)$ then

$$d_{34} = 0, d_{14} \neq 0 \text{ and } d_{24} \neq 0$$

If, in addition, Kleinman's symmetry condition is fulfilled, we obtain that for a crystal with a binary axis, there are eight nonzero coefficients and four of them are independent, defining the tensor in the following way:

$$\begin{bmatrix} 0 & 0 & 0 & d_{14} & d_{31} & 0 \\ 0 & 0 & 0 & d_{24} & d_{14} & 0 \\ d_{31} & d_{24} & d_{33} & 0 & 0 & d_{14} \end{bmatrix}$$

In this thesis the nonlinear response of materials with different symmetry will be analyzed: a quartz crystal and polymeric thin films with uniaxial and biaxial orientation. The corresponding tensors in each case will be shown in the part that is dedicated to the data treatment (Chap. 2).

1.1.4.2 SHG on a Planar Surface. Maker Fringes

As we have seen before, when an electromagnetic wave strikes a nonlinear medium a nonlinear polarization is induced, and one of its terms oscillates at the frequency 2ω. As a result of this oscillating polarization, each infinitesimal sheet with thickness Dy turns into a radiation emitter at frequency 2ω, coherent with the incident beam (Fig. 1.3). Therefore, to calculate the intensity of the beam as it exits the sheet it is necessary to introduce the sum of the harmonic waves generated at the different points in the material. In this way, an oscillating amplitude is obtained as a function of the light path in the material [1]. This is because the speed of the excitation light in the medium (frequency ω) generally differs from that of the second harmonic, so the waves generated in each element dy are out of phase with the others.

The coherence length l_c is the maximum distance along which the second harmonic waves interfere so that the resulting amplitude increases. From this point on, the intensity decreases, as is observed in Fig. 1.3b.

When the mentioned rates are equal, the sum of the waves (now in phase) gives rise to an increase of the resulting amplitude with the distance travelled (Fig. 1.3a). This is known as the phase matching condition.

Nonlinear characterization of the material, as has been seen in the previous section, lies in the determination of the coefficients in the corresponding second-order susceptibility tensor, depending on the sample symmetry. The most widely used technique to measure those coefficients is based on measurements of the so-called

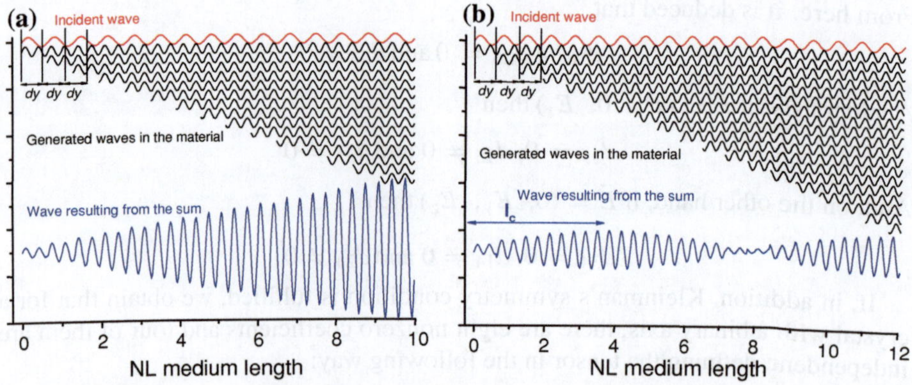

Fig. 1.3 Propagation of the waves inside the material. Schematic illustration of the phase matching of several waves. In **a** the different waves are in phase and in **b** they are not

Maker fringes. In 1962, Maker et al. [8] found that the generated and transmitted second harmonic intensity for a quartz crystal ($I^{2\omega}$) oscillated in a periodic way with the laser incidence angle. Then, nonlinear coefficients can be obtained from the amplitude of the second harmonic signal detected. The l_c can be deduced from the distance between the minima of the Maker fringes. Appendix A gives the deduction of expression 1.24, which relates second harmonic intensity at the exit of the material (measurable parameter) to the nonlinear coefficients, using the model developed by Jerphagnon and Kurtz [9]:

$$\boxed{I''_{2\omega} = \left(\frac{8\pi c}{\left((n^{\omega})^2 - (n^{2\omega})^2\right)^2}\right) d^2 p^2(\theta) \left|E'_{\omega}\right|^4 t''^4_{\omega} T''_{2\omega} \sin^2 \psi} \tag{1.24}$$

Here t'' is the air/sample transmission coefficient of the incident wave, and T_2'' the sample/air transmission coefficient of the harmonic light, both of which depend on the light polarization. The definitions of the other terms in 1.24 are as follows:

d nonlinear coefficients

$p(\theta)$ projection factor; its expression is characteristic of the configuration of the measurements, and it will be determined for each practical case

$n^{\omega}$and $n^{2\omega}$ refractive indices of the fundamental and harmonic wave, respectively

E'_{ω} fundamental wave amplitude

ψ function that depends on the light path in the medium (L) as

$$\Psi = \left(n^{\omega}\cos\theta'_{\omega} - n^{2\omega}\cos\theta'_{2\omega}\right)\left(\frac{4}{\lambda}\right)\left(\frac{\pi L}{2}\right) = \frac{\pi L}{2l_c(\theta)} \tag{1.25}$$

where l_c is the coherence length.

Second Harmonic Generation in Birefringent Materials

Throughout the deduction of 1.24, the existence of any noticeable dispersion and birefringence in the material has not been taken into account. Herman and Hayden [10] proved that for highly anisotropic systems and/or moderately dispersive ones, the JK model leads to wrong values of the nonlinear coefficients. The main differences between the two treatments concern the way they consider the reflected waves at the different surfaces when the boundary conditions are stated, and in assuming, at some points in the development of the JK model, that the nonlinear material has the same refractive index at the fundamental frequency ω as at 2ω $(n^{\omega}\theta) \cong n^{2\omega}\,\theta)$. This implies that $\theta_{\omega}' \cong \theta_{2\omega}'$, which would mean that the excitation and harmonic waves propagate in the same direction inside the material, which in fact does not happen.

The expression of the generated second harmonic power inferred in Hermann and Hayden's work is

$$\boxed{\begin{array}{c} P_{2\omega} = \dfrac{128\pi^3}{cA} d_{eff}^2 \left(\dfrac{n^{2\omega l}}{n^{2\omega}}\right)^4 \left(\dfrac{(n^{\omega})^2 - (n^{2\omega})^2}{(n^{\omega})^2 - (n^{2\omega l})^2}\right) \sin^2\psi \\ \dfrac{\left[t_{\omega}^{a\to m}\right]^4 \left[T_{2\omega}^{m\to s}\right]^2 \left[T_{2\omega}^{s\to a}\right]^2}{\cos^2\gamma_{2\omega}\cos^2(\theta_{2\omega}' - \gamma_{2\omega})(n^{\omega}\cos\theta_{\omega}' - n^{2\omega}\cos\theta_{2\omega}')^2} P_{\omega}^2 \end{array}} \tag{1.26}$$

We see here that it is essentially a function of the same parameters as JK's expression, but some differences can be observed. On the one hand, t are Fresnel coefficients (see Appendix A), and there appears a walk-off angle [11] γ, explicitly. The term "walk-off" refers to the deviation from the wave vector ($\vec{k}$) direction suffered by the energy propagation direction (given by the Poynting vector, $\vec{S}$) in an anisotropic medium, forming an angle γ (Fig. 1.4). This angle is also the one existing between the direction of the field $\boldsymbol{E}$ and the electric displacement $\boldsymbol{D}$. It only occurs for the extraordinary beam, propagating at a θ angle with respect to the optical axis, so n_{e} and the phase velocity become dependent on this angle.

Throughout the fitting of the previous expressions, it is necessary to know the refractive indices at the incident and second harmonic frequencies. Some basic concepts about light propagation through a linear medium will be given now, briefly, to establish the terminology when talking about refractive indices in the birefringent samples.

Light Propagation in a Linear Medium

The speed of light in an optically isotropic material is the same in every propagation direction and it is equal to that in the vacuum (c) divided by the refractive index $n = \sqrt{\varepsilon}$. In general terms, however, this speed depends on the direction of propagation. For any anisotropic material it is possible to find some axes (x, y, z, crystalline symmetry axes) in which the dielectric tensor ε is diagonal, and so can

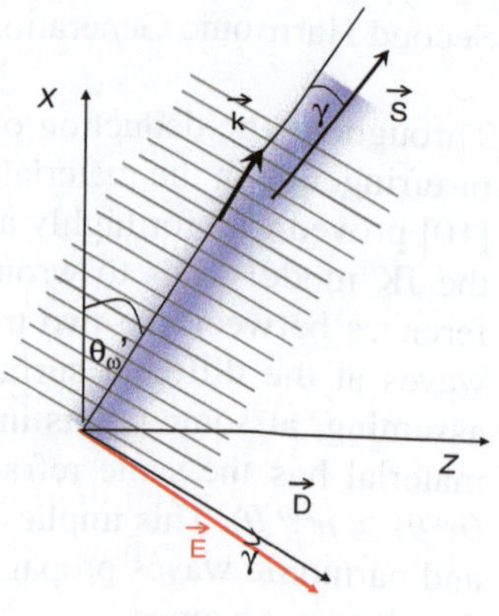

Fig. 1.4 Representation of the walk-off angle. *Grey lines* show the wavefront and the *blue zone* the region with a significant optical intensity. The angle magnitude is over dimensioned in the figure. As a direct consequence, the wave propagation efficiency is restricted [4]

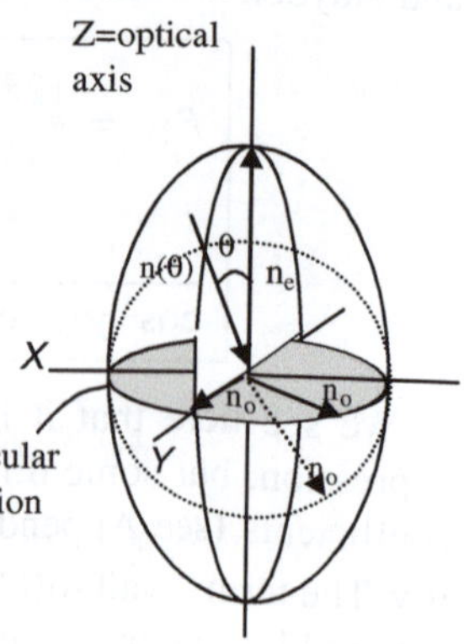

Fig. 1.5 Ellipsoid of revolution, uniaxial sample

be expressed in terms of the principal dielectric constants ε_x, ε_y and ε_z. The square roots of these constants are the principal refractive indices, and they determine the propagation rate of the polarized waves along the corresponding axes. So, a polarized wave along the x-axis propagates with a rate $v_x = c/\sqrt{\varepsilon_x} = c/n_x$, and so on [1].

Given the direction of wave propagation, its speed can be determined by the so-called refractive index ellipsoid. This is a geometric construction in which the semiaxes represent the principal values n_x, n_y, n_z, which are usually different. Depending on the symmetry, the index ellipsoid can be reduced to a sphere (isotropic medium, cubic crystalline symmetry) or to an ellipsoid of revolution (uniaxial material, a crystal with tetragonal symmetry for instance) where two of the principal indices are equal. In Fig. 1.5, an index ellipsoid is represented for a material with only one principal optical axis, the z-axis, while $n_x = n_y$.

If we focus our attention on the uniaxial case in Fig. 1.5, a wave propagating along the optical axis z, regardless of its polarization, "sees" one only index, called the ordinary index $n_o = n_x = n_y$. The unlike index, n_z in this case, is called the extraordinary index n_e, and the difference $\Delta n = n_e - n_o$ is the birefringence.

Uniaxial crystals having Δn larger than zero are called positive (as in Fig. 1.5), whereas the ones with $n_o > n_e$ are said to have negative birefringence.

In a general case, for an arbitrary direction of propagation in a uniaxial medium, the electromagnetic wave will split into two, polarized in mutually perpendicular directions, which, as we have just seen, will propagate with different velocities. The corresponding refractive indices are given by the semiaxes of the ellipse that is defined when a plane perpendicular to the propagation direction cuts the ellipsoid. One of the waves, called ordinary, will always be polarized perpendicular to the optical axis, so its index will be n_o and $v_o = c/n_o$. The other one, the extraordinary wave, will propagate with a speed given by an index that is dependent on the direction of propagation. Then, for an incidence angle θ, as shown in Fig. 1.5, the refractive index for the extraordinary wave would be expressed as

$$\frac{1}{n^2(\theta)} = \frac{\cos^2\theta}{n_o^2} + \frac{\sin^2\theta}{n_e^2} \tag{1.27}$$

This expression [12] will be utilized usually in the calculation of the refractive indices for the fundamental and harmonic waves, which is necessary for the analysis of the results of SHG in oriented polymers.

In the case of biaxial systems the axes of the ellipsoid will be different, and so $n_x \neq n_y \neq n_z$.

1.1.4.3 SHG in Molecules (EFISH)

Firstly, we are going to deal with nonlinearity from a microscopic point of view, aimed at the characterization of the second-order optical nonlinear response of organic compounds. It is considered that the molecules interact weakly, so when the global response of the group is analysed, the properties remain essentially those of the individual molecules.

One molecule, in the presence of the electric field from the incident radiation at frequency ω, behaves as an oscillating dipole, which will emit radiation at ω, 2ω, 3ω, ... The dipole moment of this molecule can be expressed as

$$\vec{p} = \alpha : \vec{E} + \beta : \vec{E}\vec{E} + \gamma : \vec{E}\vec{E}\vec{E} + \ldots \tag{1.28}$$

where α is the polarizability of the molecule and β and γ are the first- and second-order molecular hyperpolarizabilities, respectively. Although we will treat molecular design later in this chapter, it is obvious that β, the parameter that characterizes the second-order nonlinear response, is zero for centrosymmetric molecules.

Characterization at the microscopic scale is usually carried out by means of measurements in solution. However, this results in a null nonlinear macroscopic response owing to the isotropic nature of solutions, even if the individual response of the molecules is really efficient (large values of β). The EFISH technique (electric field induced second harmonic) [13–15], used in this work, is based on the breaking of the centrosymmetric of the solution by applying a static electric field that forces the molecular dipole orientation along the field direction. From the analysis of the EFISH measurements the projection of the vector part of β along the direction of the dipole moment of the molecule is determined.

If we apply the expression 1.28 to the particular case of an EFISH experiment we have

$$p_i^{2\omega} = \beta_{ijk}(-2\omega; \omega, \omega) E_j^{\omega} E_k^{\omega} + \gamma_{ijkl}(-2\omega; \omega, \omega, 0) E_j^{\omega} E_k^{\omega} E_l^{0} \tag{1.29}$$

It is interesting to note that, although the objective is to obtain a second-order molecular parameter ($\vec{\beta}$,) during the EFISH experiment a third-order phenomenon occurs, as we apply three fields to the material, two of them optical (E, E), and the third one, "static" (E^0). Then, the parameter we obtain from the EFISH measurements is the macroscopic susceptibility Γ_{ZZZZ}, which relates the polarization of the solution to the mentioned fields:

$$P_I^{2\omega} = \Gamma_{IJKL}(-2\omega; \omega, \omega, 0) E_J^{\omega} E_K^{\omega} E_L^{0} \tag{1.30}$$

In the usual configuration for an EFISH experiment the static electric field and the laser excitation field are polarized along the Z-axis, so [4]:

$$P_Z^{2\omega} = \Gamma_{ZZZZ}(-2\omega; \omega, \omega, 0) E_Z^{\omega} E_Z^{\omega} E_Z^{0} \tag{1.31}$$

The relationship between Γ_{ZZZZ} and the microscopic hyperpolarizabilities is expressed [16], as we will see in Sect. 1.1.5, as

$$\Gamma_{ZZZZ} = N f \gamma^0, \tag{1.32}$$

where γ^0 is the averaged hyperpolarizability, which includes two contributions, one second order and one third order, and which can be written as

$$\gamma^0 = \frac{\mu_0 \beta_z}{5kT} + \langle \gamma \rangle, \tag{1.33}$$

where μ_0 is the permeability of free space and k and T are Boltzmann's constant and absolute temperature, respectively. In order to calculate the exact value of β, it would be necessary to know μ and the second-order hyperpolarizability $<\gamma>$. However, for molecules with a noticeable charge transfer at room temperature, which will be the case for the chromophores studied in this thesis, μ is much bigger than $<\gamma>$, and the contribution of the latter is considered negligible.

In Appendix B, the deduction of the expression for the second harmonic intensity generated in the EFISH cell is explained, as in the case of thin films. It is a function that oscillates with the light path in the liquid, which corresponds to the typical Maker fringes record (see Fig. 2.10):

$$I_{2\omega} = 2 I_M^L \sin^2\left(\frac{\pi l}{2 l_c}\right) \tag{1.34}$$

where $I_M{}^L$ is the half-height of the Maker fringes and l_c the coherence length of the solution. From these two parameters $\mu\beta$ can be deduced.

Static Hyperpolarizability: Two-Level Model

In the calculation of the quadratic hyperpolarizabilities, it is common to use a two-level model [3, 22] to describe the dependence of β on the excitation frequency. It is based on considering that there is only one excited state that contributes in a relevant manner to the nonlinear response.

Organic molecules with π—conjugated systems, such as the ones studied in this thesis, show large second-order nonlinearity owing to the electronic delocalization in the π bonds, and the presence of electron donor and acceptor substituents in opposite positions. In these kinds of molecules there is a strong one-dimensional charge transfer, so it is possible to assume that molecule properties are determined by both states, the fundamental one and first excited state of the intramolecular chain transfer. Although this is a simplification, its use is justified in many families of compounds because the contribution to the first excited state is the most important one.

The coefficient $\beta(2\omega)$ obtained by this model is written as a function of the experimentally accessible properties, so that [17]

$$\beta(-2\omega;\omega,\omega) = \beta_Z(-2\omega;\omega,\omega) = \frac{3}{2h^2c^2}\frac{(\mu_{01})^2\Delta\mu}{(\omega_1^2-4\omega^2)}\frac{\omega_1^2}{(\omega_1^2-\omega^2)} \tag{1.35}$$

where

$\Delta\mu$ difference between the dipole moments of the two states
μ_{01} transition moment between the two states
$h\omega_1$ transition energy
$h\omega$ excitation energy

The previous expression can be separated into two factors:

$$\beta(2\omega) = \beta(0)F(\omega,\omega_1),$$

where $\beta(0)$ is the static hyperpolarizability of the molecule, which can be considered as an intrinsic measurement of the quadratic nonlinearity and $F(\omega, \omega_1)$ is the dispersion factor. $\beta(0)$ allows the nonlinear properties of the molecules to be compared regardless of their absorption spectrum.

The previous equation can be expressed as a function of wavelength λ, which is the expression we will use for the data treatment in Chap. 2:

$$\beta(0) = \beta(2\omega)\left[1-4\left(\frac{\lambda_{max}}{\lambda}\right)^2\right]\left[1-\left(\frac{\lambda_{max}}{\lambda}\right)^2\right] \tag{1.36}$$

where λ_{max} is the absorption maximum position in the UV spectrum and λ the excitation wavelength.

1.1.5 Relationship Between Microscopic and Macroscopic Nonlinearities

Once the hyperpolarizability of one molecule is known, it is possible to calculate the response of the macroscopic system of which the molecule is a part. For a crystal, where the positions and orientations are well defined, the mentioned additivity criteria are applied. In the case of a solution or a polymer oriented by an electric field, molecular orientation distribution has to be taken into account when it comes to evaluating the macroscopic response.

The group of molecules in a polymer film can be treated as an oriented rigid gas, so the relationship between molecular hyperpolarizabilities and the macroscopic nonlinear coefficients is well defined. This relationship was developed by Meredith et al. (1982–1984) and Williams et al. and Singer et al. in 1987 [18–21].

The susceptibility $\chi^{(2)}_{IJK}$ is related to the microscopic coefficients by

$$\chi^{(2)}_{IJK} = Nf(0)f(\omega)^2 f(2\omega)b_{IJK}, \tag{1.37}$$

where N is the number of nonlinear molecules, f includes the different local field factors and b_{IJK} represents the β_{ijk} tensor, written in the macroscopic reference system.

The local field factors relate the applied external field $E(\omega)$ to the field that the molecules perceive in the condensed phase. For highly symmetric media the Lorentz correction factor is employed [2]:

$$f(\omega) = \frac{(n^{\omega})^2 + 2}{3} \text{ and } f(2\omega) = \frac{(n^{2\omega})^2 + 2}{3}, \tag{1.38}$$

where n^{ω} and $n^{2\omega}$ are the refractive indices at frequencies ω and 2ω.

This factor takes account of the effects of dipoles induced by the electronic polarization in the medium. Moreover, the external fields and the fields associated with the neighbouring dipoles can induce orientation of the permanent dipoles, which implies an additional correction in the f factor. Local field factors ascribed to this process, known as Onsager's factor, are added. They are written as [3]

$$f(0) = \frac{\varepsilon_0\left((n^{\omega})^2 + 2\right)}{(n^{\omega})^2 + 2\varepsilon_0}, \tag{1.39}$$

where ε_0 is the static dielectric constant.

The tensor b_{IJK} shows the same features as the tensor β_{ijk}, but it is expressed in the film's reference system. It describes the contribution of a single molecule to the macroscopic nonlinear coefficients, and it is related to β_{ijk} as

$$b_{IJK} = \sum_{ijk}\left(\cos\theta_{Ii}\cos\theta_{Jj}\cos\theta_{Kk}\right)\beta_{ijk} \tag{1.40}$$

Here, θ_{Ii} is the angle between the macroscopic system axis *I*, and the molecular system axis *i*. The term cos θ_{Ii} is the scalar product of the unit vectors *I* and *i* and it gives the projection of the molecular coefficients on the crystallographic reference system.

The problem can be simplified remarkably by assuming that the predominant microscopic component in the second order tensor (β_{zzz}) is parallel to the dipole moment μ, as happens in our case. Besides, considering electric-field-oriented polymeric systems, the chromophore dipoles tend to align in the direction of the applied field, *Z*. In this situation, the only relevant angle is θ, between the molecular axis and the macroscopic axis *Z*. When evaluating b_{IJK} it has to be taken into account that we have an orientation distribution of the molecules around the polar axis *Z*, so the orientation averages $<\cos\theta>$ need to be calculated. In these conditions, as shown in the data treatment in Chap. 2, there are only two nonzero components of the *b* tensor:

$$<b_{ZZZ}>=<\cos^3\theta>\beta_{zzz}, \tag{1.41}$$

$$<b_{ZXX}>=\frac{<\cos\theta-\cos^3\theta>}{2}\beta_{zzz} \tag{1.42}$$

So the relevant elements in the susceptibility tensor will be

$$\chi_{ZZZ}=Nf<\cos^3\theta>\beta_{zzz}, \tag{1.43}$$

$$\chi_{ZXX}=Nf\frac{<\cos\theta-\cos^3\theta>}{2}\beta_{zzz}, \tag{1.44}$$

where

$$f=f(0)f(\omega)^2f(2\omega)$$

The calculation of the mentioned orientation averages is performed by considering an orientation distribution function, $F(\theta)=\exp(-U(\theta)/kT$, where U($\theta$) is the energy of the molecule in the presence of the field. As detailed in Appendix C, in this way, the expression for the nonlinear coefficients d_{ij} for an initially isotropic medium can be deduced as a function of the microscopic hyperpolarizability, in the limit for low fields and dipole moments [4]

$$d_{33}=1/2\chi^{(2)}_{ZZZ}=Nf\frac{\mu_0\beta_Z}{10kT}E^0, \tag{1.45}$$

$$d_{31}=1/2\chi^{(2)}_{ZXX}=Nf\frac{\mu_0\beta_Z}{30kT}E^0, \tag{1.46}$$

where k is the Boltzmann constant, $k=1.38\times10^{-16}$ esu $=1.38\times10^{-23}$ JK^{-1}, T the temperature and E_0 the amplitude of the orienting field.

1.2 Materials for Nonlinear Optics

1.2.1 Introduction

Any material may generate a second harmonic, as long as it lacks a centre of symmetry. However, in practice, some requirements need to be fulfilled. These are a large nonlinearity and a fast response, a wide spectral range of transparency, good processability (which allows its incorporation into a device) and good stability with respect to thermal or photo-oxidative degradation.

At first, the search for NLO materials was focused on inorganic materials but it was extended soon after to organic ones as well. In the 1970s, Davidov et al. [22] proved that organic molecules with donor and acceptor groups might be suitable systems for SHG. Since then, a vast number of compounds for NLO have been designed and synthesized, while pursuing the above-mentioned properties. The advantages and drawbacks versus inorganic crystals are gathered in Table 1.1.

As a first approximation, an added feature of the organic materials is that it is possible to predict their behaviour in the bulk from the determination of the nonlinear response at the molecular level characterized by the hyperpolarizability β.

1.2.2 Organic Materials for SHG. Donor–π–Acceptor Chromophores

It has already been mentioned several times in this chapter that the necessary condition for SHG is the absence of a centre of inversion. At the molecular scale, this is achieved by using polarizable structures in an asymmetric way. The origin of the nonlinearity in the organic compounds lies in the high mobility or electron delocalization in the π bonds, particularly in the conjugated systems (compounds in which simple and multiple bonds appear alternately).

Traditionally, molecules with nonzero dipole moment in the fundamental state have been studied, that is, dipolar molecules with an electron donor group on one end of the conjugated structure and an acceptor one on the other end (push–pull systems). These kinds of push–pull molecules have a large dipole moment in the fundamental state, and show an intense transition in the UV–vis spectral region, assigned to the intramolecular charge transfer (CT) along the molecular axis. A simple molecule corresponding to this scheme, *p*-nitroaniline, is shown in Fig. 1.6 [23].

In these systems the first CT excited state is considered to control the optical nonlinear response. As has been explained in previous sections, by using the two-level model proposed by Oudar and Chemla [24], NLO properties can be expressed in terms of the transition moments (or oscillator strengths), excitation energies and dipole moment differences:

$$\beta \propto \frac{\Delta\mu_{01}\,\mu_{01}^2}{E^2} \propto \frac{\Delta\mu_{01}\,f_{01}}{E^3} \qquad (1.47)$$

Table 1.1 Advantages and disadvantages of the use of organic materials for NLO

Advantages	Disadvantages
Great versatility in their synthesis to modify their molecular structure	Low transparency in UV–vis; this restricts the operating range and decreases the optical damage threshold
Larger non-resonant optical nonlinearities	Thermal and photochemical instability with increasing conjugation of the compound
Less response time due to the π bonds	Worse mechanic properties
Smaller dielectric constants (electro-optic applications)	
Easy processability	

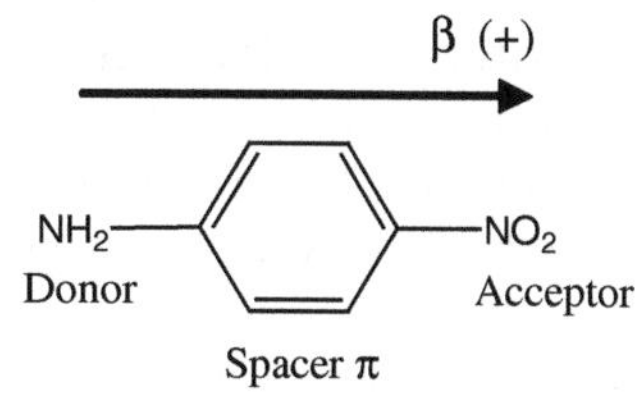

Fig. 1.6 *p*-Nitroaniline structure

$\Delta\mu_{01}$ is the difference between the dipole moment in the first excited state and in the fundamental one. E is the difference in energy between the two states. μ_{01} is the transition dipole moment, which is connected to the oscillator strength f_{01} as $f_{01} \propto \mu_{01}{}^2E$. The oscillator strength f_{01} is related to the area under the absorption curve in the UV–vis spectrum.

This model was very useful, at first, for designing chromophores with the optimum nonlinear response. The effects of changing the strength of the donor–acceptor groups and the π-conjugated system length on the NLO response were studied in detail. When the relative "strength" between the two groups is increased, larger β values are achieved, but the CT transition undergoes a red shift (bathochromic displacement), decreasing the range of transparency. On the other hand, generally, a larger conjugated chain (until a saturation point is reached) gives rise to an increase in β [25].

Since the 1990s, new criteria have been incorporated in the strategies to control NLO molecular properties. Thus, Marder et al. [26–30]. suggested that the use of compounds with an aromatic character in their fundamental state (and a basically quinoid excited state) was not suitable for optimizing β, highlighting the importance of the polarization degree in the fundamental state of the molecule, and relating it directly to β. From the molecular design point of view, it is helpful to describe the fundamental and first excited states as a combination of two extreme resonance structures: the neutral one (Fig. 1.7a) and the zwitterionic or charge-separated form (Fig. 1.7b). The relative contribution of these forms to the fundamental state determines its polarization degree.

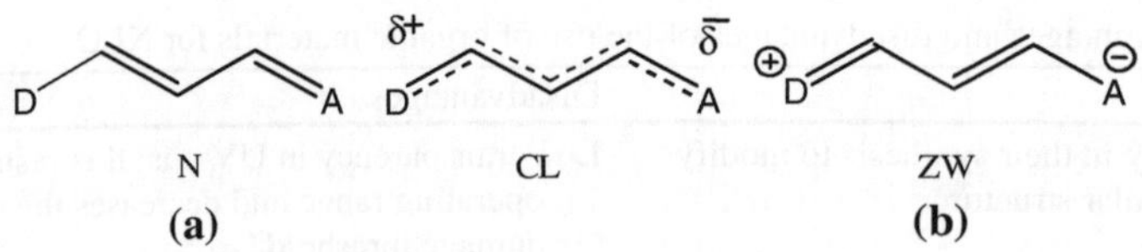

Fig. 1.7 Representation of the resonance forms of the donor–π–acceptor molecules. **a** neutral form **b** resonance or zwitterionic form

In order to have a measurement of this polarization degree a structural parameter called bond length alternation (BLA) was introduced. It is defined as the average difference in length between single and double bonds in the polyenic chain. It is affected by the chemical structure (strength and nature of the donor and acceptor groups), the spacer topology and the surroundings of the molecule itself (solvent polarity). This parameter can be evaluated from the experimental data obtained by nuclear magnetic resonance (NMR) or theoretically calculated.

Between the two resonance forms there is an intermediate one that corresponds to the so-called cyanine limit (CL), in which both extreme forms contribute equally.

In the case of a polyene substituted by weak donor and acceptor groups, the form that better represents the fundamental state would be (a), showing a large value of BLA. If we increase the strength in both groups, the (b) resonance form would have a higher contribution and, therefore BLA would decrease. The relationship between BLA and the hyperpolarizability β is shown in Fig. 1.8. As can be observed in the figure, when the charge-separated contribution dominates the description of the fundamental state, BLA and β are negative.

As a strategy in the molecular design, the incorporation of proaromatic fragments in the donor groups as well as in the acceptor ones was proposed, so that the molecules gain aromaticity in the charge transfer, stabilizing the zwitterionic form and thus increasing the β value. Several studies were able to prove how the β value can be optimized, either positive or negative, by varying BLA. This is achieved by modifying donor and acceptor substituents and the π-conjugated bridge [26, 31–36].

1.2.2.1 Solvent Influence

The development of theoretical models and computational methods in order to describe the effect of the solvent on NLO properties of molecular systems face great difficulties, as the nature of solute–solvent interactions is extremely complicated.

Before explaining the influence on the nonlinear response, we will briefly comment on the solvatochromism phenomenon, which is described as the change in position, intensity and shape of the absorption UV–vis band of the chromophore in solvents with different polarity [37]. Only the solute–solvent

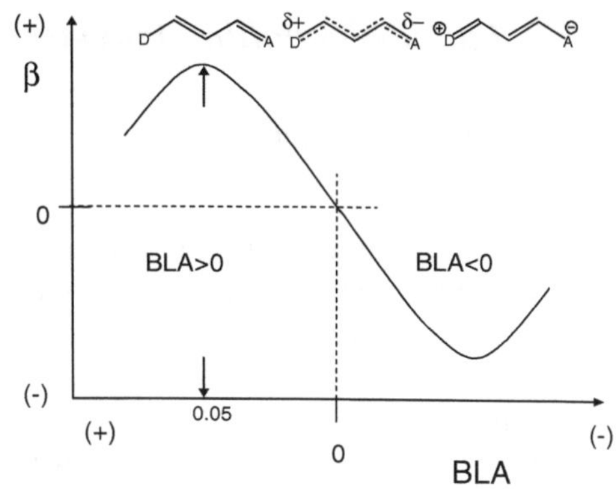

Fig. 1.8 Marder's curve representing nonlinearity versus *BLA* parameter

interactions will be considered here, leaving aside the solute–solute type ones. In spite of the fact that the observed shifts in the absorption spectrum have their origin in several types of interactions [38], we will only comment on the electrostatic ones.

The electronic transition from the fundamental state to the CT excited one entails an important change in the electron density distribution of the molecule, so a large change in the permanent dipole moment is observed during the excitation. Owing to the difference in solute–solvent electrostatic interaction energy between the fundamental and the CT states, larger solvatochromic shifts are observed as the solvents become more polar. Thus, there exists a linear dependence between solvatochromic shift and the difference between dipole moments in the fundamental and excited states. Positive solvatochromism (bathochromic or red shift) becomes evident in molecules with larger polarization for the CT state than for the fundamental one. For these kinds of molecules, very polar solvents bring more stability to the excited state, so the excitation energy decreases significantly. In negative solvatochromism (hypsochromic or blue or shift), polar solvents stabilize the fundamental state more, with a corresponding increase in the transition energy.

1.2.2.2 Solvatochromism and Molecular Hyperpolarizability

The origin of the changes in the NLO response of the donor–π–acceptor chromophores as a function of the solvent polarity is related again to the solute–solvent interactions in the fundamental and CT excited states.

Molecules showing positive solvatochromism are characterized by a larger β as the solvent polarity increases, whereas for molecules showing negative solvatochromism, β decreases as the polarity of the solvent increases. Therefore, it is possible that the same compound may have positive and negative hyperpolarizability, as it will be assigned to one zone or another of Marder's curve depending on the solvent in which it is measured.

1.2.3 Liquid-Crystalline Polymers with Azobenzene

1.2.3.1 Liquid Crystals

The liquid-crystalline state can be considered as intermediate between solid, showing perfect long-range order, and liquid, which is isotropic and fluid. The molecules in a liquid crystal (LC) are not totally ordered in space, but they show certain orientational fluidity. LCs can be classified depending on the way the liquid-crystalline phase is induced. Thus, there are lyotropic liquid crystals, in which the transition is induced by increasing the concentration in the solution, and thermotropic liquid crystals, in which it is induced thermally. Thermotropic mesogens (compounds showing liquid-crystalline phases depending on the temperature) show a large variety of shapes. They can be elongated (calamitic), disk-shaped (discotic) or bent-core (banana). In all of them, the mesogen axis tends to align along a direction or a director axis [39].

Depending on the spatial arrangement of the molecules, different phases can be distinguished. For calamitic liquid crystals the mesophases can be as follows.

Nematic (N): The mesogens have no positional order, but they self-align to have long-range directional order with their long axes parallel, the positions of their centres of mass being randomly distributed (Fig. 1.9).

Smectic (Sm): The mesogens form well-defined layers, without any long-range order inside each layer. They show translational periodicity in one dimension. There are several types of smectic phases. In Smectic A, Sm_A, the molecules are oriented along the layer normal (Fig. 1.9).

A single material can show different phases depending on the temperature, ranging from most ordered at low temperatures to least ordered at high temperatures.

1.2.3.2 Polymers

Polymeric materials are suitable for a great variety of applications because of the versatility in their synthesis and processing as well as their relatively low cost of production. Polymers are macromolecules made up of many repeating basic units called monomers and they can be composed of one single monomeric structure—giving rise to homopolymers—or several,. If they are composed of two types of monomers they are called copolymers. Among copolymers we can find several types depending on the monomer order: random, block or graft.

All the macromolecules in a polymeric material do not have the same size, so the molecular weight of a polymer is described by two parameters, the weight-average molecular weight M_w and the number-average molecular weight M_n. The first represents the average of the molecular masses of the individual

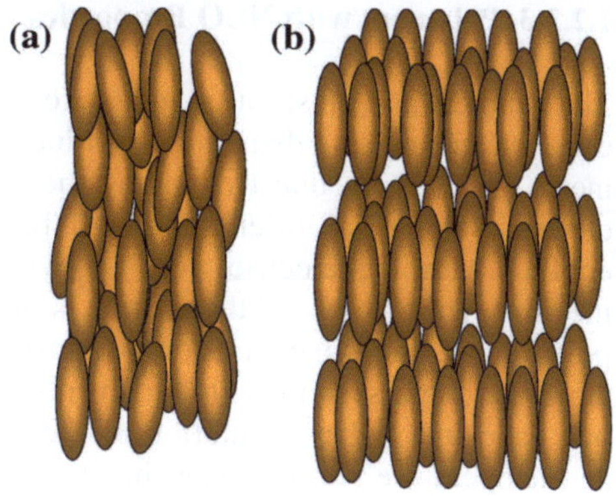

Fig. 1.9 Schematic representation of a mesophase: **a** Nematic (N) and **b** Smectic A (SmA)

macromolecules, and the second, the weighted average of the masses per each molecule's mass. The ratio M_w/M_n is the polydispersity index, and it is equal to or greater than one. The mechanical properties (tensile strength, hardness, etc.-) depend on the intermolecular forces along the polymer chain, so they improve to some extent when the molecular weight is increased, depending on the polymer type. However, the viscosity of the polymer is influenced by the entanglements formed by the polymeric chains. As for the polymer architecture, the most common structure is the linear one, but there are polymers with side chains, branched randomly or subsequently (dendrimers), or star-shaped. The semicrystalline state is the most common one in the polymers. The chains intermingle, showing amorphous or disordered zones and certain regions where they are ordered, forming crystals. In semicrystalline polymers there are two types of transitions, one arising from the crystalline part and the other from the amorphous part. On reaching the glass transition temperature T_g (second-order transition) the amorphous part, which is rigid and disordered at low temperatures, begins to show fluidity and, at higher temperatures is found to be in the liquid phase. The other transition, a first-order one, is caused by the crystalline part, which, after reaching the melting temperature T_m, is found to be in a more or less viscous liquid state. The values of T_m are normally 33–100 % higher than Tg [40]. In general terms, it can be said that both T_g and T_m increase with the stiffness of the chain. The stronger the molecular interactions and the size of the rigid groups (aromatic, conjugated double bonds, etc.) the larger the rigidity. The chain symmetry increases T_m but decreases T_g. If a polymer is cooled down so quickly from its state at $T > T_m$ to temperatures below T_g (quenching) that it does not have enough time to order into crystals it can be "frozen" in an amorphous state.

The glass transition occurs at a higher temperature as the cooling rate increases. This is because a temperature is reached at which the conformational movements, required by the chain to reach equilibrium, are slower than the experimental time scale, reaching a metastable or "frozen" state that relaxes more slowly the lower the temperature is [41].

1.2.3.3 Polymers with NLO Properties

The simplest and most intuitive way to prepare polymer films with large nonlinear coefficients is by dispersing nonlinear molecules with high β values in polymeric matrices, so that by heating the film above T_g and applying an external electric field, polar orientation will be forced in the molecules. Subsequent cooling with the electric field on will give rise to a non-centrosymmetric medium. The χ value of these films is then a function of the obtained degree of polar order, the concentration of active molecules and the magnitude of β. However, these systems exhibit several problems that have to be solved. On the one hand, the polar orientation of the chromophores tends to relax with time and temperature. On the other hand, as the polarity of the molecules increases they tend to be less soluble in the polymeric chain, so it is not possible to achieve high chromophore contents, because phase separation or crystallization in microdomains occur. Chemical bonding of the chromophores to the polymeric chain allows an increase in their concentration and, moreover, improves the stability of the polar order induced by poling, owing to the restriction in the chromophore movement as it is linked to the polymer. In spite of the very extensive work performed in this field, many of the most relevant results come from Dalton et al. [42]. Their most recent strategies for improving nonlinear properties are based on the design of molecular and polymeric structures hindering dipole–dipole intermolecular attractive forces to avoid aggregation. This group has designed systems with coefficients r_{33} ($r_{33} = -4d_{33}/n^4$) [43] between 300 and 500 pm V^{-1} (at $\lambda = 1.3$ µm), by incorporating the chromophore dispersed in dendritic structures containing the same chromophore and aryl fluoroaryl groups [3] [44] .

1.2.3.4 Liquid-Crystalline Polymers with NLO Properties

Liquid crystalline polymers (LCPs) are polymeric structures having mesogenic units inserted in the principal chain or linked to it by a more or less flexible spacer

[3] These systems are called BCOGs (binary chromophore organic glasses). An example of a dispersed chromophore is YLD_124:

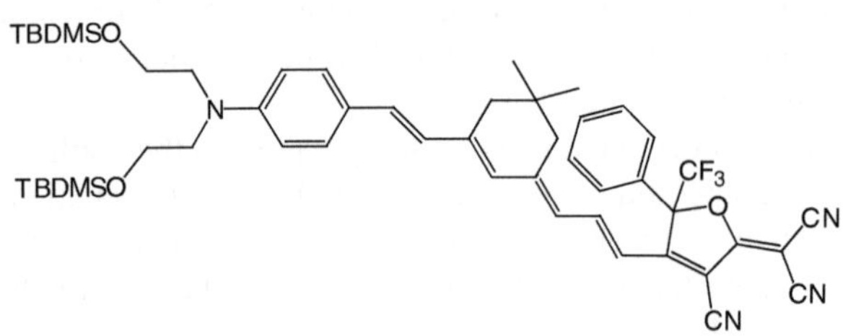

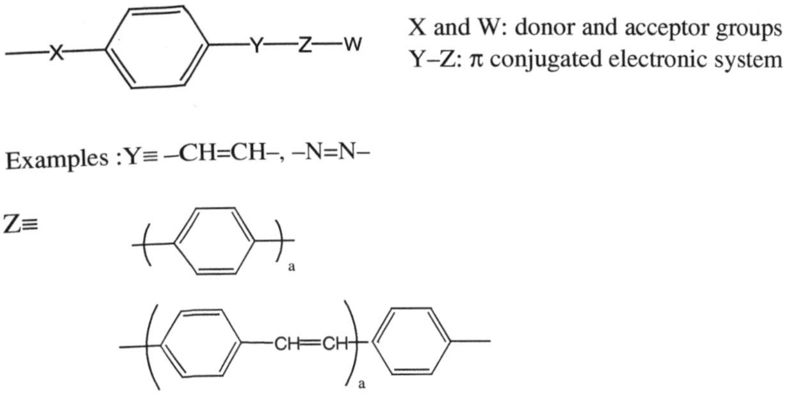

Fig. 1.10 Donor–π–acceptor-type mesogen structures

as side chains. In this study, all the employed LCPs are thermotropic and show calamitic shape.

These kinds of structures are very versatile as they offer the possibility of changing different parameters affecting the mesophase formation. An increase in the flexibility in the principal chain or in a side chain (by lengthening the methylenic spacer) decreases T_g and favours mesophase formation. Short spacers favour nematic phases, while long ones favour smectic ones. The mesogen ordering in the side chain can decouple from that in the polymer principal chain because of the flexibility of this spacer, thus allowing mesophase formation.

Usually, there is no crystallization in LCPs, and below the liquid crystalline phase the glass transition is found. The orientational order obtained in one mesophase can be "frozen" in by simply cooling the polymer below its T_g. However, if it is suddenly cooled down from the isotropic phase to temperatures well below T_g (quenching), the disordered state remains "frozen" in.

Owing to the anisotropic nature of the mesogens, LCPs show anisotropy in their optical properties, which becomes clear in the birefringence and dichroism measured in the films. Birefringence is the difference between refractive indices for perpendicular directions of light polarization, as has been explained in this chapter, while dichroism is the difference in the optical absorption of the polarized light for different directions.

The first to explore the possibility of using side-chain liquid-crystalline polymers as a nonlinear medium were Meredith et al. in 1982 [18–21]. Some of the key factors for NLO activity in the molecules (high axial ratio, planarity) are in line with those promoting liquid-crystalline properties in low molecular weight compounds. Therefore, a great effort has been made to synthesize liquid crystalline polymers with NLO properties in which nonlinear moieties show mesogenic properties by themselves. Some of the investigated structures are schematically depicted in Fig. 1.10 [43].

Fig. 1.11 Disperse red 1

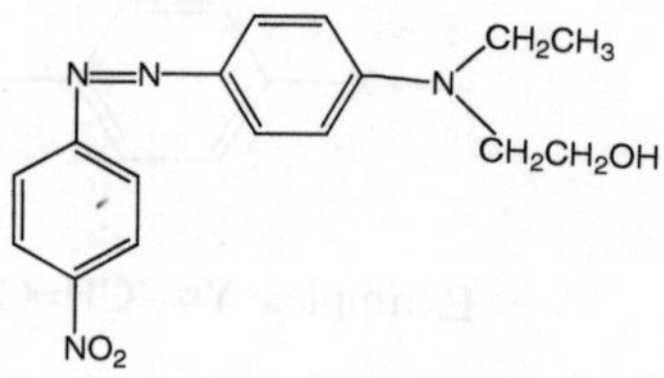

The interest in exploring the nonlinear response in liquid-crystalline polymers has been due, partly, to the fact that it was expected that the axial order, characteristic of systems with nematic or smectic phases, could contribute to favouring the polar order. However, the obtained results have been very diverse, and in some cases even contradictory. Just to mention some examples, Worboys et al. [45]. prepared a number of copolyacrylates containing Disperse Red 1 as a side chain in different concentrations, giving rise to amorphous or liquid-crystalline materials depending on this chromophore content. Disperse Red 1 (DR1) shown in Fig. 1.11 is a typical nonlinear azochromophore often used as a standard for EFISH measurements at 1.9 μm. In their work, they found more intensity and stability of the nonlinear parameters in the case of liquid crystals.

In a later study Koide et al. [46]. found that in the case of liquid crystals the ratio χ_{33}/χ_{31} was much larger than 3, which is the theoretical value predicted for isotropic systems, confirming that a larger anisotropy in the nonlinear response is achieved in the case of LCs (see Appendix C). This fact was observed by several authors, as is reported in the review by Dubois et al. [43]. However, other authors state that it is impossible to achieve efficient polar orientations in the case of LCPs with smectic phases [47, 48]. This would be related to the fact that the obtained axial order in a liquid crystal favours the intermolecular interactions between chromophores, which can decrease the intensity of the nonlinear response.

In any case, these interactions often also entail an increase in the stability of the obtained order versus the characteristic anisotropy of the amorphous materials, which is beneficial for the stability of the generated nonlinear response.

The copolymers based on mesogenic and nonlinear side chains offer the opportunity to change the properties of the polymer precisely, by modifying the mesogen content.

1.2.3.5 Azobenzene

The chromophores in the side-chain LCPs whose NLO properties are studied in this thesis have the azobenzene group as the π-conjugated system. The azopolymers have proved to be excellent materials for NLO [49–52]. By incorporating the azo system, an added advantage is available, since light's influence upon these chromophores can be used, as will be discussed later.

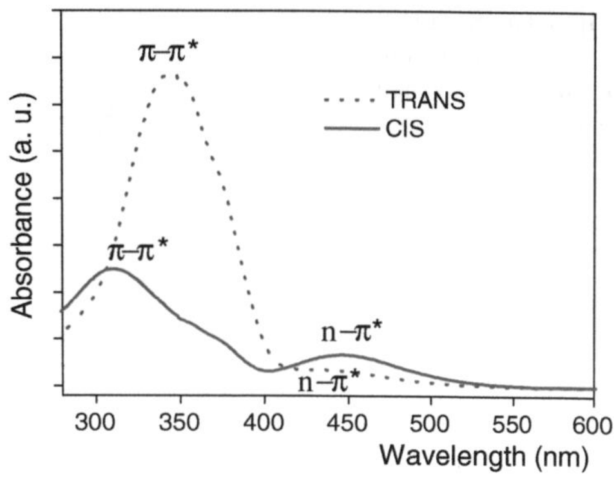

Fig. 1.12 Absorption spectrum of a polymethylmethacrylate derivative with a side chain containing 4-cyano-4-alkoxy azobenzene

The interest in the azobenzene group is essentially based on the *trans–cis–trans* isomerization [53] (Fig. 1.12), which occurs in response to light irradiation with the suitable wavelength. The molecule, usually in *trans* "*E*" conformation, can change to *cis* "*Z*" by UV–vis irradiation. These isomers, having different geometry and dipole moment, show different absorption spectra. The *trans* isomer shows two absorption bands. One of them, of low intensity and located in the visible region, is associated with $n-\pi^*$ transition, which is actually a result of nonplanar distortions in the molecule and vibrational couplings, as this transition is forbidden in planar symmetry. The other band, with higher intensity, is assigned to the $\pi-\pi^*$ transition and it is situated in the near UV. When the isomerization occurs, the $\pi-\pi^*$ band is observed shifted to higher energies and with less intensity than that for the *trans*, while the $n-\pi^*$ band becomes more intense, keeping its position more or less.

The *cis*–to-*trans* isomer conversion (thermodynamically more stable) can happen by thermal or photochemical means, while the *trans* to *cis* conversion only happens by light irradiation.

In the scientific literature, mainly two possible mechanisms of photoisomerization are discussed: rotation and inversion. The rotation mechanism is based on a turn around the $-N = N-$ bond out of the molecule's plane. On the other hand, in the inversion mechanism the occurs a turn of one or both of the nitrogens in the molecule's plane through a stationary sp hybrid state, as shown in Fig. 1.13.

The particular characteristics of the isomerization undergone by the molecule depend on the types of substituents in *ortho* and *para* positions that are added to the aromatic rings. Thus, H. Rau [54] established a classification into three groups for azobenzoic compounds according to the position of the $\pi-\pi^*$ and $n-\pi^*$ transitions: azobenzene-type, aminoazobenzene-type, and pseudostilbene.

Azobenzene-type compounds show an absorption spectrum like that in Fig. 1.12, in which the n-π^* transition appears at lower energies than the $\pi-\pi^*$ one, while the so-called *aminoazobenzene* compounds show both bands very near to each other. This is because the introduction of donor substituents in the

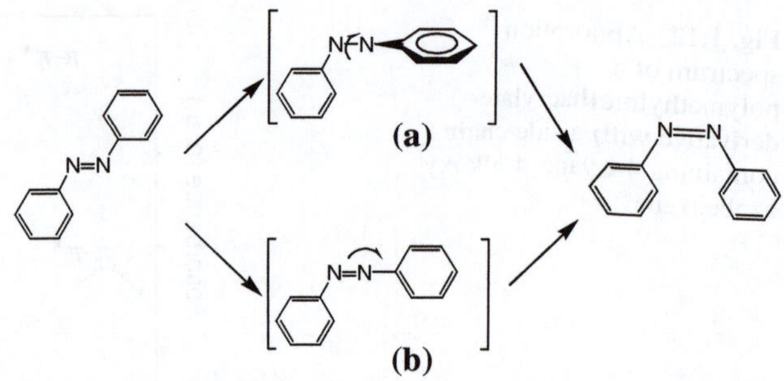

Fig. 1.13 Rotation (**a**) and inversion, (**b**) mechanisms for the azobenzene *trans–cis* photosiomerization

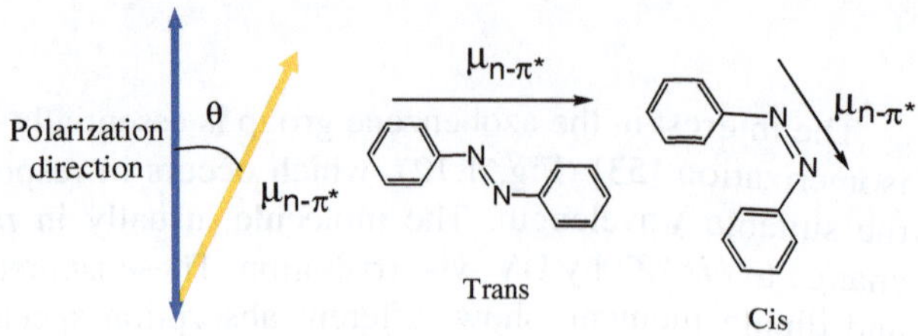

Fig. 1.14 Direction of the μ dipole corresponding to the n$-\pi^*$ electron transition for the *trans* and *cis* isomers

ortho and *para* positions increases the π orbital's energy and decreases that of the π* orbital. Finally, in the *pseudoestilbene* group the $\pi-\pi^*$ band undergoes such a bathochromic shift that the positions of the two bands can be reversed. This is what happens in the 4 and 4′ azobenzenes substituted by donor and acceptor groups. When the donor or acceptor character of these groups is increased, the half-life of the *cis* isomers decreases [55].

One of the consequences of the isomerization processes is the so-called Weigert effect, consisting of the induced orientation in the azobenzene molecules by linearly polarized UV-blue light. The probability of linearly polarized light absorption is proportional to $\cos^2\theta$, where θ is the angle between the direction of the molecular transition dipole $\mu_{\mathrm{n}-\pi^*}$ and that of the light polarization, as shown in Fig. 1.14 [56].

If we have an isotropic distribution of *trans* isomers, those with an orientation near to the direction of polarization will absorb preferably. After passing through the *cis* state, they will go back to *trans* with a different orientation than the initial one. A *trans* "falling" in the direction perpendicular to the light polarization stays in this situation.

After successive cycles of isomerization, azobenzene molecules are left oriented preferably in a plane that is perpendicular to the excitation light polarization, as shown in Fig. 1.15.

Larger values and better stability of the photoinduced anisotropy have been found in mesogenic systems with azobenzene units than in amorphous systems. This behaviour has been attributed to the cooperative effects characteristic of liquid-crystalline compounds [57, 58].

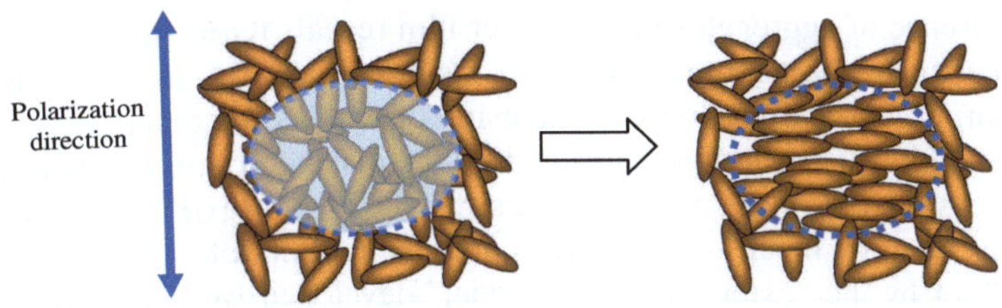

Fig. 1.15 Photoinduced order in compounds with azobenzene units

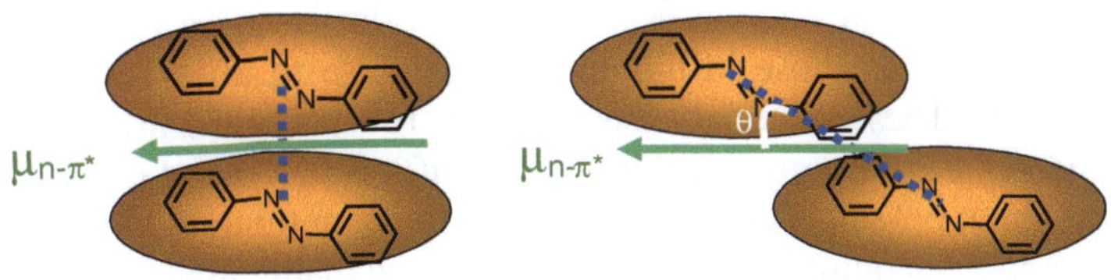

Fig. 1.16 Representation of the two different types of azobenzene aggregates, J and H

The use of NLO chromophores containing azobenzene groups in their structure is very useful, as it is possible to use light irradiation instead of heating in order to orient them by means of the electric field. In this way, polar orientation is achieved at temperatures well below the polymer T_g. This process is called photoassisted poling (PAP) and it consists in taking advantage of the free volume generated by the isomerization of the moieties *trans* to *cis*, to orient the molecules by means of the electric field, because it increases their mobility at temperatures below T_g. This method was first applied in poly (methylmethacrylate)/DR1 (PMMA/DR1) with the DR1 dispersed or chemically bonded to the main chain [59, 60], but it has been extended to different systems containing azobenzene[50, 51].

The aggregation of azobenzenes is the stacking of the aromatic rings, which align parallel to the electronic transition dipoles. The chromophore–chromophore interactions giving rise to these aggregations are van der Waals type, and take place mainly between chromophores comprising an extended π-conjugated system, as in the case of azobenzene. This may happen in solution as well as in the solid state, and it is a function of both the chromophore structure (donor and acceptor character of the substituents, conjugated system length) and the environment surrounding the molecules. Among the described different types of chromophores, we can find H-type aggregates (Fig. 1.16), which are one-dimensional molecular systems where the transition moments align perpendicularly to the line connecting the molecular centres, and J-type aggregates, in which the transition moments of each azobenzene align such that they form an angle θ with the line connecting the molecular centres.

The presence of aggregates in a polymer film reveals itself in its optical properties. In particular, the aggregation causes a substantial shift in the peak of maximum absorption. The direction and extension of this displacement depends on the number of aggregated chromophores and on the distance between them inside the aggregate, as well as on its orientation, that is, the angle θ [61]. This phenomenon can be described by means of the "molecular exciton model", described by Kasha et al. [62], or by the "extended dipole model" developed by Kuhn and coworkers [63]. J-type aggregation, named after Edwin E. Jelley (one of the first researchers who investigated this phenomenon), displaces the absorption band from the position occupied by the isolated chromophore to higher wavelengths, while H-type aggregates cause a hypsochromic shift.

The aggregation of polymers containing donor–π–acceptor azobenzene chromophores, such as those in this thesis, gives rise to absorption spectra with very wide bands. These materials usually show H-type aggregation more clearly, with the line in the spectrum shifted to the blue, if it is compared to the spectrum of the isolated chromophore. These shifts are due to the antiparallel arrangement that the chromophores acquire, which is more thermodynamically favoured than the parallel stacking, as the chromophores usually have high dipole moments. It is considered that the architecture of these aggregates is very diverse, thus anything from dimers to aggregates can be found, with different sizes and packing geometries. According to the exciton model, the interactions between nearby chromophores give rise to electron transitions to states with different oscillator strengths [64], so the absorption spectrum provides information about the presence of the mentioned aggregates but it does not allow a quantitative analysis of the supramolecular architecture [65]. Nevertheless, Menzel et al. [66] suggested a semi-quantitative procedure to calculate the relative quantities of the aggregated azochromophores and the isolated ones. By means of a deconvolution of the UV–vis absorption bands, the complex absorption band decomposes into three bands, corresponding to the two types of aggregation and the isolated azobenzenes. From the relative area under the corresponding curves an estimate of the fraction of each moiety can be obtained.

References

1. J. M. Cabrera., F. Agulló., F. J. López, Óptica electromagnética vol. 2, Addison-Wesley Iberoamericana Española, Madrid (2000)
2. R.W. Boyd, *Nonlinear Optics* (Academic Press, San Diego, 1992)
3. R.L. Sutherland, *Handbook of Nonlinear Optics* (Marcel Dekker, New York, 1996)
4. P.N. Prasad, D.J. Williams, *Introduction to Nonlinear Optical Effects in Molecules and Polymers* (Wiley, New York, 1991)
5. P.A. Franken et al., Generation of optical harmonics. Phys. Rev. Lett. **7**(4), 118 (1961)
6. T. Verbiest, K. Clays, V. Rodriguez, *Second-Order Nonlinear Optical Characterization Techniques* (Taylor & Francis, Boca Raton, FL, 2009)
7. D.A. Kleinman, Theory of second harmonic generation of light. Phys. Rev. **128**(4), 1761 (1962)
8. P.D. Maker et al., Effects of dispersion and focusing on the production of optical harmonics. Phys. Rev. Lett. **8**(1), 21 (1962)

9. J. Jerphagnon, S.K. Kurtz, Maker fringes: a detailed comparison of theory and experiment for isotropic and uniaxial crystals. J. Appl. Phys. **41**(4), 1667 (1970)
10. W.N. Herman, L.M. Hayden, Maker fringes revisited: second-harmonic generation from birefringent or absorbing materials. J. Optical Soc. Am. B **12**(3), 416 (1995)
11. R. Paschotta, *Encyclopedia of Laser Physics and Technology*, vol. 2 (Wiley-VCH, Weinhiem, Germany, 2008)
12. J. Casas Peláez, *Óptica* (Librería Central, Zaragoza, 1994)
13. B.F. Levine et al., Molecular hyperpolarizabilities determined from conjugated and nonconjugated organic liquids. Appl. Phys. Lett. **24**(9), 445 (1974)
14. J.L. Oudar, H. Le Person, Second-order polarizabilities of some aromatic molecules. Optics Commun **15**(2), 258 (1975)
15. K.D. Singer et al., Measurements of molecular second order optical susceptibilities using dc induced second harmonic generation. J. Chem. Phys. **75**(7), 3572 (1981)
16. I. Ledoux, J. Zyss, Influence of the molecular environment in solution measurements of the second-order optical susceptibility for urea and derivatives. Chem. Phys. **73**(1–2), 203 (1982)
17. S. Di Bella et al., Coordination and organometallic complexes as second-order nonlinear optical molecular materials. Top. Organomet. Chem. **28**, 1 (2010)
18. G.R. Meredith, J. VanDusen, D.J. Williams, Optical and nonlinear optical characterization of molecularly doped thermotropic liquid crystalline polymers. Macromolecules **15**(5), 1385 (1982)
19. G. R. Meredith, J. G. VanDusen, D. J. Williams, in "*Nonlinear Optical Properties of Organic and Polymeric Materials*", ACS Symposium Series, vol. 233, ed. by D. J. Williams, ACS, Washington, DC, Chap. 5, pp. 109–133 (1983)
20. K. D. Singer et al., in "*Nonlinear Optical Properties of Organic Molecules and Crystals*", Vol. 1, ed. by D. S. Chemla and J. Zyss (1987), Academic, New York, p. 437
21. D. J. Williams et al., in "*Nonlinear Optical Properties of Organic Molecules and Crystals*", vol. 1, ed. by D. S. Chemla and J. Zyss, Academic, New York, p. 405 (1987)
22. B.L. Davydov et al., Connection between charge transfer and laser second harmonic generation. J. Exp. Theor. Phys. Lett **12**, 16 (1970)
23. H.S. Nalwa, S. Miyata (eds.), *Nonlinear Optics of Organic Molecules and Polymers* (CRC, Boca Ratón, FL, 1997)
24. J.L. Oudar, D.S. Chemla, Hyperpolarizabilities of the nitroanilines and their relations to the excited state dipole moment. J. Chem. Phys. **66**(6), 2664 (1977)
25. R.A. Huijts, G.L.J. Hesselink, Length dependence of the second-order polarizability in conjugated organic molecules. Chem. Phys. Lett. **156**(2–3), 209 (1989)
26. S.R. Marder, D.N. Beratan, L.-T. Cheng, Approaches for optimizing the first electronic hyperpolarizability of conjugated organic molecules. Science **252**(5002), 103 (1991)
27. S.R. Marder et al., Relation between bond-length alternation and second electronic hyperpolarizability of conjugated organic molecules. Science **261**(5118), 186 (1993)
28. F. Meyers et al., Electric field modulated nonlinear optical properties of donor-acceptor polyenes: sum-over-states investigation of the relationship between molecular polarizabilities (α, β, and γ) and bond length alternation. J. Am. Chem. Soc. **116**(23), 10703 (1994)
29. S.R. Marder et al., A unified description of linear and nonlinear polarization in organic polymethine dyes. Science **265**(5172), 632 (1994)
30. M. Ahlheim et al., Chromophores with strong heterocyclic acceptors: a poled polymer with a large electro-optic coefficient. Science **271**(5247), 335 (1996)
31. S.R. Marder et al., Large first hyperpolarizabilities in push-pull polyenes by tuning of the bond length alternation and aromaticity. Science **263**(5146), 511 (1994)
32. S.R. Marder et al., Design and synthesis of chromophores and polymers for electro-optic and photorefractive applications. Nature **388**(6645), 845 (1997)
33. E. Aqad et al., Novel D-π-A chromophores based on the fulvene accepting moiety. Org. Lett. **3**(15), 2329 (2001)
34. C.B. Gorman, S.R. Marder, An investigation of the interrelationships between linear and nonlinear polarizabilities and bond-length alternation in conjugated organic molecules. Proc. Nat. Acad. Sci. **90**(23), 11297 (1993)

35. G. Bourhill et al., Experimental demonstration of the dependence of the first hyperpolarizability of donor-acceptor-substituted polyenes on the ground-state polarization and bond length alternation. J. Am. Chem. Soc. **116**(6), 2619 (1994)
36. C.B. Gorman, S.R. Marder, Effect of molecular polarization on bond-length alternation, linear polarizability, first and second hyperpolarizability in donor-acceptor polyenes as a function of chain length. Chem. Mater. **7**(1), 215 (1995)
37. C. Reichardt, Solvatochromic dyes as solvent polarity indicators. Chem. Rev. **94**(8), 2319 (1994)
38. J. Li, C.J. Cramer, D.G. Truhlar, Two-response-time model based on CM2/INDO/S2 electrostatic potentials for the dielectric polarization component of solvatochromic shifts on vertical excitation energies. Int. J. Quantum Chem. **77**(1), 264 (2000)
39. F. J. Rodríguez, Propiedades ópticas fotoinducidas en polímeros con uidades de azobenceno Ph.D. Dissertation, Universidad de Zargoza (2005)
40. R.B. Seymour, C.E. Carraher, *Polymer Chemistry: An Introduction* (CRC Press, Boca Raton, FL, 1988)
41. J. Areizaga et al., "*Polímeros*" Síntesis, Madrid (2002)
42. L.R. Dalton, P.A. Sullivan, D.H. Bale, Electric field poled organic electro-optic materials: state of the art and future prospects. Chem. Rev. **110**(1), 25 (2010)
43. J.C. Dubois et al., Behavior and properties of side chain thermotropic liquid crystal polymers. Acta. Polym. **48**(3), 47 (1997)
44. L.R. Dalton et al., Guest–host cooperativity in organic materials greatly enhances the nonlinear optical response. J. Phys. Chem. C **112**(11), 4355 (2008)
45. M. R. Worboys, et al., in *Proceedings 2nd International Conference on Electrical, Optical and Acoustic Properties of Polymers* Institute of Materials, London (18-1, 18-6) (1990)
46. N. Koide et al., Thermal transition behavior and second nonlinear optical properties of polymers containing mesogenic side chains. Mol. Cryst. Liq. Cryst. **198**, 323 (1991)
47. I. Rau, F. Kajzar, New insights into the relaxation of polar order in electro-optic polymers. Thin Solid Films **516**(24), 8880 (2008)
48. F. Kajzar, C. Noël, Molecular design and properties of side chain liquid crystal polymers for applications in optoelectronics. Adv. Mater. Opt. Electron. **8**(5), 247 (1998)
49. D.M. Burland, R.D. Miller, C.A. Walsh, Second-order nonlinearity in poled-polymer systems. Chem. Rev. **94**(1), 31 (1994)
50. L.R. Dalton et al., Synthesis and processing of improved organic second-order nonlinear optical materials for applications in photonics. Chem. Mater. **7**(6), 1060 (1995)
51. J.A. Delaire, K. Nakatani, Linear and nonlinear optical properties of photochromic molecules and materials. Chem. Rev. **100**(5), 1817 (2000)
52. S.K. Yesodha, C.K. Sadashiva Pillai, N. Tsutsumi, Stable polymeric materials for nonlinear optics: a review based on azobenzene systems. Prog. Poly. Sci. **29**(1), 45 (2004)
53. G.S. Hartley, The cis-form of azobenzene. Nature **140**(3537), 281 (1937)
54. H. Rau, in "*Photochemistry and Photophysics*" vol. 12, ed. by J. Rebek CRC Press, Boca Ratón, FL (1990)
55. T. Bieringer, in "*Holographic Data Storage*" ed. by H. J. Coufal, D. Psaltis and G. T. Sincerbox, Springer, New York, p. 209 (2000)
56. T. Huang, K.H. Wagner, Diffraction analysis of photoanisotropic holography: an anisotropic saturation model. J. Optical Soc. Am. B **13**(2), 282 (1996)
57. N.C.R. Holme, P.S. Ramanujam, S. Hvilsted, Photoinduced anisotropy measurements in liquid-crystalline azobenzene side-chain polyesters. Appl. Opt. **35**(23), 4622 (1996)
58. C. Sánchez et al., Biphotonic holographic gratings in azobenzene polyesters: surface relief phenomena and polarization effects. Appl. Phys. Lett. **77**(10), 1440 (2000)
59. Z. Sekkat, M. Dumont, Photoassisted poling of azo dye doped polymeric films at room temperature. Appl. Phys. B **54**(5), 486 (1992)
60. P.M. Blanchard, G.R. Mitchell, A comparison of photoinduced poling and thermal poling of azo-dye-doped polymer films for second order nonlinear optical applications. Appl. Phys. Lett. **63**(15), 2038 (1993)

61. H. Menzel, in "*Photorefractive Organic Thin Films*", ed. by Z. Sekkat and W. Knoll Academic Press, San Diego (2002)
62. M. Kasha, H.R. Rawls, M. Ashraf El-Bayoumi, The exciton model in molecular spectroscopy. Pure Appl. Chem. **11**(3–4), 371 (1965)
63. V. Czikklely, Kuhn H. Försterling, Extended dipole model for aggregates of dye molecules. Chem. Phys. Lett. **6**(3), 207 (1970)
64. A.M. Kelley, A multimode vibronic treatment of absorption, resonance Raman, and hyper-Rayleigh scattering of excitonically coupled molecular dimers. J. Chem. Phys. **119**(6), 3320 (2003)
65. A. Priimägi, "*Polymer-Azobenzene Complexes: From Supramolecular Concepts to Efficient Photoresponsive Polymers*" Ph.D. Dissertation, Helsinki University of Technology (2009)
66. H. Menzel et al., Small-angle X-ray scattering and ultraviolet-visible spectroscopy studies on the structure and structural changes in langmuir-blodgett films of polyglutamates with azobenzene moieties tethered by alkyl spacers of different length. Langmuir **10**(6), 1926 (1994)

References

61. H. Menzel, in "*Photorefractive Organic Thin Films*", ed. by Z. Sekkat and W. Knoll (Academic Press, San Diego, 2002)
62. M. Kasha, H.R. Rawls, M. Ashraf El-Bayoumi, The exciton model in molecular spectroscopy. Pure Appl. Chem. 11(3-4), 371 (1965)
63. V. Czikkely, H.D. Försterling, H. Kuhn, Extended dipole model for aggregates of dye molecules. Chem. Phys. Lett. 6(3), 207 (1970)
64. A.M. Kelley, A multimode vibronic treatment of absorption, resonance Raman, and hyper-Rayleigh scattering of excitonically coupled molecular dimers. J. Chem. Phys. 119(6), 3320 (2003)
65. A. Priimagi, *Polymer-Azobenzene Complexes: From Supramolecular Concepts to Efficient Photoresponsive Polymers*. Ph.D. Dissertation, Helsinki University of Technology (2009)
66. H. Menzel et al., Small-angle X-ray scattering and ultraviolet-visible spectroscopy studies on the structure and structural changes in Langmuir-Blodgett films of polyglutamates with azobenzene moieties tethered by alkyl spacers of different length. Langmuir 10(6), 1926 (1994)

Chapter 2
Experimental Methods

The techniques and equipment used for the characterization of the materials (at the molecular and macroscopic level) that have been used in this thesis are detailed below. Besides, the procedures used to modify the properties of the polymers of the study, mainly based on optical and thermal treatments, are also explained.

2.1 Sample Preparation

2.1.1 Synthesis and Basic Characterization of the Materials

The Department of Organic Chemistry of the University of Zaragoza took care of the synthesis of the molecules and polymers used in this thesis. Prof. Garín's group synthesized the highly efficient NLO chromophores whose hyperpolarizability in solution were measured as explained in Chap. 3, whereas the Liquid Crystal and Polymer group was responsible for the monomer and polymer synthesis of the research. Both groups belong to the Institute of Materials Science of Aragon (ICMA—Instituto de Ciencia de Materiales de Aragón).

The chemical structure of the synthesized compounds was confirmed by infrared spectroscopy techniques and ^{1}H NMR. The particular features of its mesomorphic nature, as well as the thermal properties, were studied by means of differential scanning calorimetry and electronic microscopy. The molecular weights were determined by gel permeation chromatography (GPC).

The monomeric units of the main polymer of this research, Pol-PZ-CN (see Fig. 4.1), were prepared by azo-coupling of *N*-(hydroxyhexyl)-*N′*-phenylpiperazine with the 4-cyanoaniline diazonium salt. The radical polymerizations, thermally initiated by AIBN (azobis(isobutyronitrile)), were carried out in DMF (dimethylformamide) kept at a temperature of 70 °C [1].

R. Alicante, *Photoinduced Modifications of the Nonlinear Optical Response in Liquid Crystalline Azopolymers*, Springer Theses, DOI: 10.1007/978-3-642-31756-9_2,

Fig. 2.1 Azopolymer films and mixtures of polymers with chromophores prepared by free evaporation of CH_2Cl_2

2.1.2 Film Preparation and Thickness Measurement

The polymer films were produced by "casting" followed by free evaporation of the solvent. A small amount of the polymer (1–3 mg) was dissolved in 0.8 ml of filtered dichloromethane (CH_2Cl_2) and placed on three glass or fused silica substrates, with a surface of 2×2 cm^2. The thickness and quality of the films obtained in this way depend on the concentration of the solution and the rate of evaporation of the solvent, among other things, so several concentrations were tested until transparent and homogeneous films of about 0.5 μm thickness were obtained for some studies and in the range of 1–2 μm for others.

In order to produce the non-centrosymmetric orientation needed for second harmonic signal generation, a corona discharge on the sample was used. The substrates used in these measurements had one of their faces coated with a $\cong 100$ Å thick layer of ITO (indium tin oxide), which is a transparent conductor. Fused silica substrates (Qu-Suprasil) are also used for the optical absorption and refractive index measurements.

Substrate cleaning is a major necessity in order to obtain homogeneous films of good optical quality. The substrates underwent cleaning by sonication in several steps, following a protocol in which distilled water, soap and finally methanol were used. After that, they were subjected to UV irradiation in an ozone photoreactor. This method removes, by oxidation, any organic residues that may be on the surface, such as water or traces of solvent, grease from the skin or simply pollutants that might have stuck during prolonged exposure to air.

Once the glass was clean, the solution deposition of the concerned polymer was performed. In order to remove possible traces of solvent, the films were kept covered at 40 °C on a hotplate for 12 h. In Fig. 2.1, a group of films that were part of the research in this thesis are shown. These films were prepared on glass substrates with ITO, as has been explained, and are composed of azopolymers, PMMA and mixtures of these with donor–π–acceptor chromophores.

It is necessary to know the thickness of the films in order to calculate the second-order nonlinear optical coefficients. To obtain the thickness two independent methods were used. One of the measurements was obtained using a contact profilometer Veeco Dektak^{3}ST, which also allowed the surface relief of the samples to be measured, with a vertical nominal resolution in the range of nanometres along a straight line of up to 50 mm length. In practice, in good quality films, up to 5 nm can be discerned. Previous to the measurement, a groove was made in the film with a bradawl, thus crossing through its thickness to get to the other surface of the substrate while avoiding notching it. The measurement was carried out using a needle with a 2.5 μm radius diamond point, which scanned the surface film. The irregularities of the sample make the needle move vertically, sending an electric sign to the differential transformer that is coupled to it. In these measurements, the strength that the point makes on the film can be controlled in order not to damage it. In our case, we worked with the minimum one, 10 mg.

2.2 Optical Measurements

In this optical measurements section, the basic characterization techniques of the linear optical properties, as well as other techniques more specifically used to obtain the parameters quantifying the nonlinear response at the molecular and macroscopic scale of the studied materials, are included.

2.2.1 Optical Absorption Spectroscopy (UV–vis-NIR)

As a preliminary step to the measurements of SHG, not only in the chromophores in solution but also in the polymer films, it is necessary to know the absorption spectrum. This limits, for example, the wavelengths that can be used in the NLO measurements. Besides, in the case of films, the absorption measurements give information on the molecule ordering in the films, degree of aggregation, etc.

2.2.1.1 Theoretical Basis

By using optical absorption spectroscopy, the transitions between electronic states involving the external electrons of a system are studied. In the case of organic compounds, the UV and visible radiation absorption is restricted to certain functional groups, with valence electrons of low excitation energy.

These transitions entail a change in the electronic distribution, which leads to relatively wide absorption bands. In Fig. 2.2, a scheme of the different types of transitions occurring between the π, σ and n molecular orbitals is shown.

Among the schematic transitions in Fig. 2.2, the $n \rightarrow \pi^*$ and $\pi \rightarrow \pi^*$ ones constitute the basis of the absorption spectroscopy of organic compounds. These transitions need an unsaturated group in the molecule to deliver π electrons. The

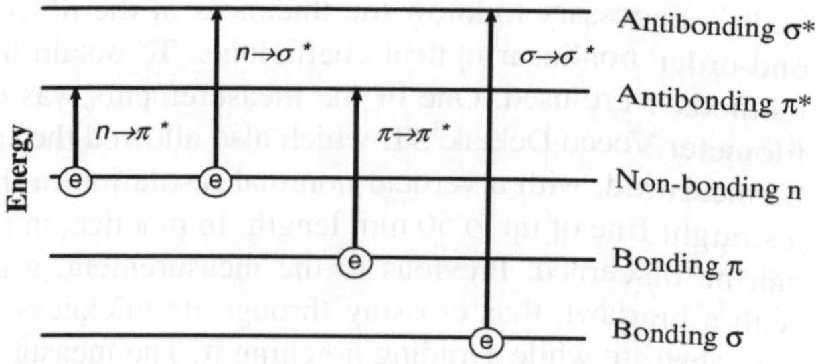

Fig. 2.2 Electronic transitions for π, σ and n electrons

maxima for these transitions fall in the region of the visible spectrum, between 400 and 700 nm.

Solvent Effect. The energy of the schematized orbitals in Fig. 2.2 changes with the polarity of the medium. Therefore, the absorption spectrum of the molecules depends on the solvent in which it is measured. If we focus on the $\pi - \pi^*$ transitions, the interaction between the solvated molecule and the solvent is translated into a decrease of the energy in the fundamental and excited levels, which is usually clearer in the last one, so a shift of the corresponding band to a lower energy (a bathochromic shift, i.e. to longer wavelengths) can be seen, giving rise to positive solvatochromism, as explained in Chap. 1. These interactions, between molecule and solvent, increase with the polarity of the latter.

Absorption coefficient. When a solution reaches an intensity I_0, the intensity transmitted can be expressed as $I_t = I_0 \times 10^{-OD}$, where OD is the absorbance of the sample. The absorbance or optical density, A, is directly proportional to the optical path l and the concentration of the absorbing species c. The Beer–Lambert law states that

$$A = \varepsilon c l,$$

where ε is a constant of proportionality called absorptivity or absorption coefficient. During this work, for some compounds, it was necessary to calculate this absorption coefficient at different wavelengths, to make the relevant absorption corrections in the $\mu\beta$ values experimentally obtained by EFISH.

2.2.1.2 Experimental Setup

The optical absorption measurements were performed with a Varian Cary-500 UV–vis-NIR spectrophotometer, which measures the intensity I_t of the light transmitted by the sample as a function of the wavelength. It is a double beam spectrophotometer, which uses a tungsten halogen lamp as an exciter for wavelengths over 350 nm and a deuterium arc lamp for the UV. The wavelength is selected using a double monochromator with 0.05 nm resolution. The beam is divided into two, one reference beam and one that crosses the sample, and then the respective light intensities transmitted by the sample and the reference are compared.

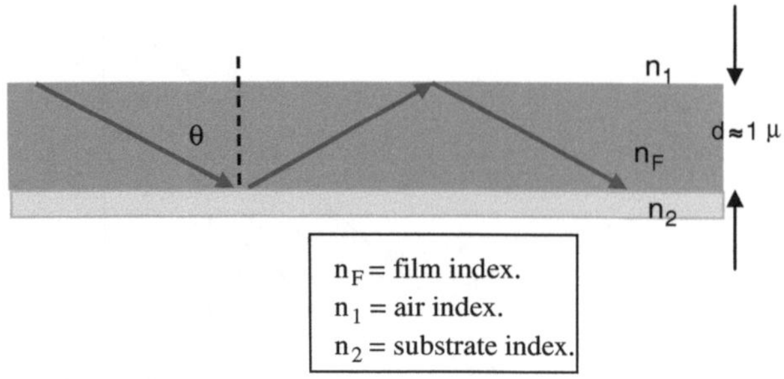

Fig. 2.3 Representation of light propagation in a plane waveguide (polymer film)

Detection in the UV–vis range is by means of a photomultiplier and in the near IR with a PbS detector.

The wavelengths that were used with the polymer films ranged from 200 to 800 nm, except for the mixtures, in which the range extended up to 1200 nm. If ITO glass substrates are used, only wavelengths down to 300 nm can be reached, as the glass absorbs a lot under that value. For the particular case of some solutions of nonlinear chromophores, the range was also increased to frequencies in the near IR.

The measurements were made with non-polarized light and linearly polarized light. For the measurements with linearly polarized light, a Glan–Thompson prism was placed in the beam of the sample. On the other hand, the incorporation of a mount that could be rotated around a vertical axis allowed measurements to be made at an angle different to normal incidence.

2.2.2 *Refractive Index Measurement*

It is necessary to know the values of the refractive indices of the films at 1.91 μm and 953 nm for the further calculation of the nonlinear coefficients. Besides, knowing the n_x, n_y, and n_z refractive indices allows us to characterize the chromophore ordering in the film.

2.2.2.1 Theoretical Basis

In order to measure the refractive indices and the thickness of the polymer films, a technique based on light guiding with the film acting as a waveguide is used. The method is based on the fact that the light can be confined in a narrow transverse region with a refractive index higher than that of its surrounding medium and then guided through the medium by consecutive internal total reflections at its boundaries (Fig. 2.3). For a given wavelength of the incident light, thickness and index of the film, the light is only guided for certain values of the angle θ (propagating modes).

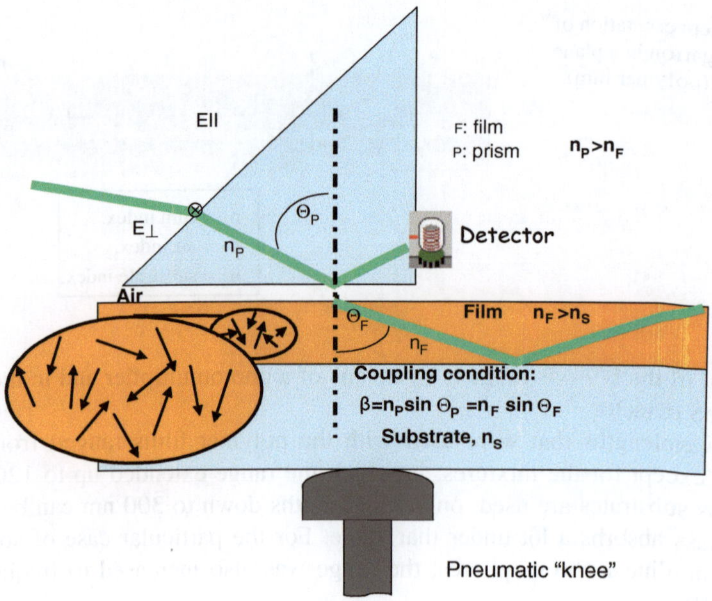

Fig. 2.4 Scheme of the experimental light prism coupling system in the film to measure the refractive indices

The propagating modes are classified as TE (transverse electric) or TM (transverse magnetic), depending on whether the incident light is polarized perpendicular or parallel to the incident plane. Once the angles at which the light is guided for the two polarizations are known, it is possible to deduce the values of the refractive indices: n_{TE} for the light polarized in the direction parallel to the plane of the film and n_{TM} perpendicular to it. For example, in the case of electric-field-oriented polymeric uniaxial films, where a Z optical axis is defined (Fig. 2.6) perpendicular to the film plane, the ordinary index $n_o = n_x = n_y$ will be measured as n_{TE} and the extraordinary one, $n_e = n_z$, will be deduced from the magnetic guided modes.

2.2.2.2 Experimental Setup

The refractive indices were measured using Metricon 2010 equipment, which includes a prism-coupled guided modes instrument.

The film, on a glass or quartz substrate, is placed in contact with the base of a prism through a pneumatic "knee", whose force on the film can be controlled by varying the pressure of the compressed air (Fig. 2.4). Nevertheless, sometimes it is necessary to increase that pressure, so for this purpose a mechanic clamp that is controlled manually was designed. Once the light coupling into the film has been achieved by using the right pressure and correct focusing of the laser beam, the process is completely automated. The film/prism arrangement is rotated so that the angle of the incident light is changed and the detected intensity is registered as a function of the angle.

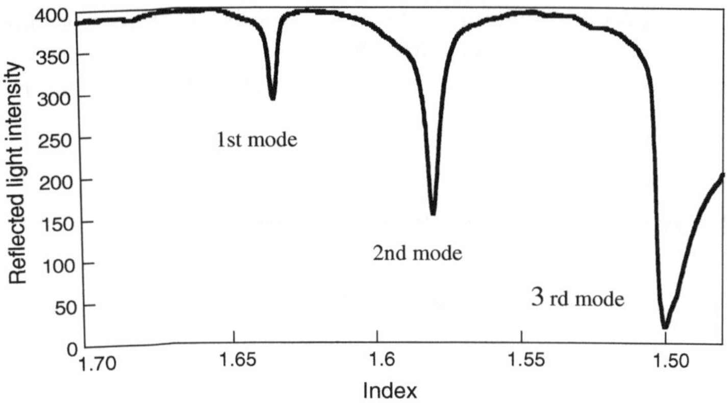

Fig. 2.5 Reductions in light intensity detected occur for angles at which the coupling condition is fulfilled

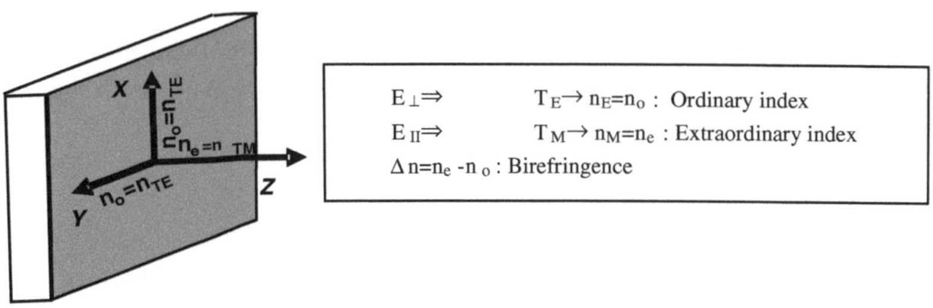

Fig. 2.6 Definition of the ordinary and extraordinary indices with respect to the film axes

The laser beam, after striking the prism and reflecting completely at the surface of the film, reaches the detector except for certain discrete values of the incidence angle, where the propagating modes are located. For these values of the angle, the light penetrates the film in wave guiding propagation mode. When this happens, that is, when the coupling condition is fulfilled, the detector registers an intense decrease of the light intensity reaching it. Figure 2.5 shows the typical guided modes record, which corresponds to the rough decrease of the light intensity that reaches the detector.

As soon as the modes are found, the equipment calculates the index and the thickness using an algorithm [2]. The angle at which the first mode is located and the angular separation between the modes allow the index and thickness of the film to be obtained. The coupling of two modes is therefore enough to determine uniquely the indices and the thickness.

The out-of-plane birefringence of the films, which will be referred to throughout this thesis, is the difference $\Delta n = n_{\mathrm{TM}} - n_{\mathrm{TE}}$. Therefore, n_{o} is the index in any direction in the film plane and n_{e} in the normal direction, as shown in Fig. 2.6.

The number of modes that can be guided increases with the film thickness. Films of at least 100–200 nm thickness are needed in most film/substrate combinations, so that the first mode can be seen at 633 nm. As has already been mentioned, two peaks are needed to determine the index and the thickness, so the minimum thickness will be ≅ 500 nm, given the substrates used.

The indices of the films were measured at the following wavelengths: 633, 780, and 1306 nm.

2.3 Nonlinear Response Characterization

2.3.1 Electric Field Induced Second Harmonic Generation

In this work, the NLO properties have been studied at the microscopic level for a large number of chromophores. It is a key characterization because the molecular properties are responsible for the nonlinear macroscopic response when the NLO chromophores are incorporated in the polymer systems. Besides, from a basic point of view, these measurements allow the results of calculations based on different theoretical models to be checked in the framework of NLO structure–properties relationship research. The theoretical background of these measurements is explained in Chap. 1, in the section dedicated to EFISH, as well as in Appendix B.

2.3.1.1 Experimental Setup

EFISH allows us to determine $\vec{\mu_0}\,\vec{\beta}$, which is the scalar product of the molecule's dipole moment in the fundamental state with the vector part of the first-order hyperpolarizability tensor $\tilde{\beta}(-2\omega,\omega,\omega)$. The Maker fringes are obtained by measuring the second harmonic light generated by a solution where a static electric field is applied, and which moves in a perpendicular direction to the incident beam. $\vec{\mu_0}\,\vec{\beta}$ is deduced from the intensity and spacing of these fringes.

The experimental setup for the SHG measurements in this work is shown schematically in Fig. 2.7. The setup, as can be seen in the figure, consists of (a) a light source, (b) a NLO spectrometer and (c) a high-voltage generator, which are described in more detail below.

(a) Light source

The fundamental beam is generated by a Nd:YAG pulsed laser, Quantel "YG780 Q-switched", which emits at wavelength $\lambda = 1064$ nm with a pulse duration of 8 ns and a maximum energy of 1 J/pulse. The of the laser beam, measured energy at the laser exit, used to make the measurements in this thesis was about 2 mJ/pulse. The frequency of repetition of the pumping lamps is 10 Hz.

In order to get light at 1.9 μm, the Nd:YAG beam is focused through a Raman cell, which consists of a 50 cm length tube in which high pressure hydrogen is locked (30 bar in this case). The stimulated Raman effect produces a shift in the incident beam frequency as a result of the vibrational mode excitation of the H_2

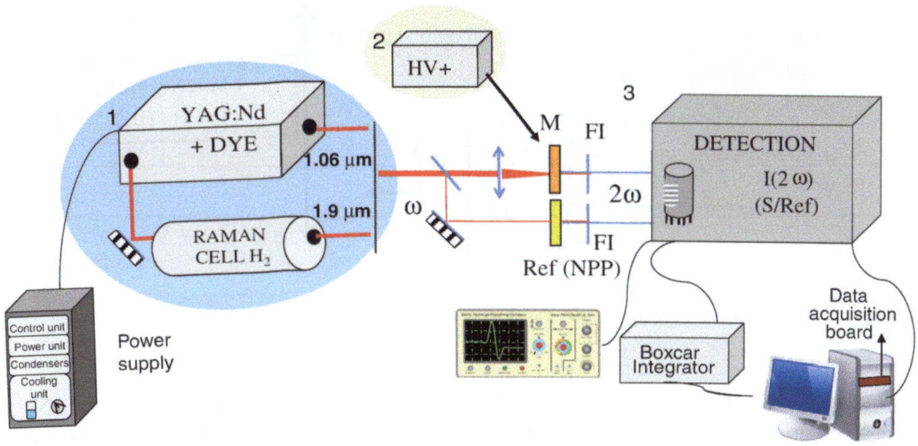

Fig. 2.7 Scheme of the SHG measurement equipment

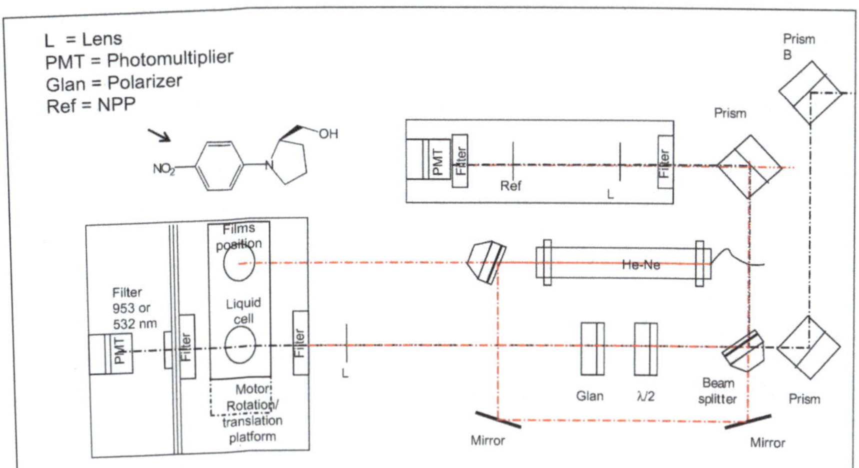

Fig. 2.8 Scheme of the sopra spectrometer used

molecule. If the incident frequency is ω, the result is light of frequencies $\omega + n\Delta\omega$ (anti-Stokes lines) and $\omega - n\Delta\omega$ (Stokes lines), where n is a whole number and $\Delta\omega = 4160\ \text{cm}^{-1}$ the H_2 Stokes shift. The first Stokes line for pumping at 1.06 μm will be 1.91 μm.

(b) Nonlinear optics spectrometer

(b1) Optical elements and detection system

The spectrometer used was purchased from Sopra. As shown in Fig. 2.8, the excitation light enters the spectrometer and, by means of two prisms, is conducted to

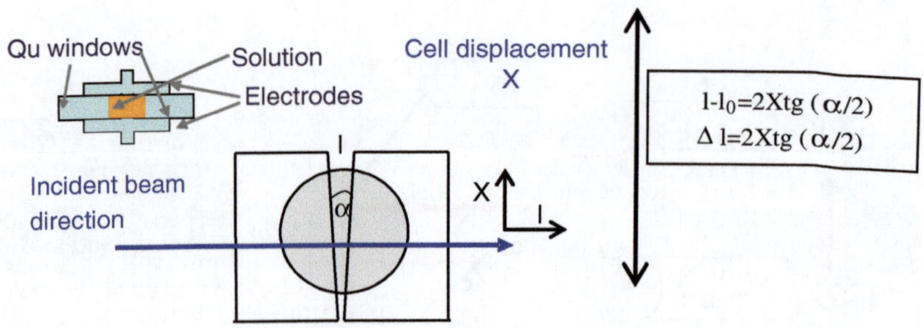

Fig. 2.9 Scheme of the cell for liquid measurements

a beam splitter, which directs approximately 10 % of the light intensity to a NPP (*N*-(4-nitrophenyl)-L-prolinol) sample. Suitable neutral filters are placed in front of the NPP sample as well as the lens that focuses the beam on it. The NPP is a non-centrosymmetric crystalline powder whose second harmonic signal is used as a reference to correct fluctuations of the light intensity of the laser. The intensity of the harmonic signal coming from the solution is divided by that obtained for the NPP. The rest of the light is directed towards the solution to be studied, passing through a half-wave (λ/2) plate, which allows us to turn the angle of the light polarization plane to adjust its intensity, and a Glan–Thompson polarizer, which determines the polarization of the excitation light (P^{ω}, parallel to the incident plane, which in our case is horizontal, and S^{ω}, perpendicular). For the EFISH measurements, the excitation light has to be vertically polarized, since the signal is much less intense for P^{ω}. The light is centred using a focusing lens in the cell containing the solution that we want to measure. The light coming from the sample passes through suitable interference filters (532 or 953 nm, full width at half maximum (FWHM) 20 nm) before reaching the detector.

In order to centre the sample as well as the reference, a He–Ne laser is used, at 633 nm. The He–Ne beam follows the same optical path as the excitation light of 1.06 or 1.91 μm.

The second harmonic generated by the sample and by the reference is detected using photomultipliers (Hamamatsu R2949 and R406), whose signal is integrated and later processed by the acquisition board (Keithley DAS 1600) incorporated in the computer.

(b2) Measurement cell

The cell where the solution to be measured is placed consists of two wedge-shaped quartz windows (2 × 1 × 0.2 cm^3) forming a relatively small angle (2–4° for measurements at 1.06 μm and 5–7° for measurements at 1.91 μm). The solution to be measured is then contained in the wedge-shaped space, and limited above and below by two electrodes (Fig. 2.9). The whole system is placed on a motorized platform, which moves in the direction perpendicular to the incident beam.

In order to determine the molecular hyperpolarizability of a compound in solution, it is necessary to know the solvent's hyperpolarizability, as it is a relative measurement. Therefore, in the experimental protocol we follow, the generated signal of the solvent is measured before and after each solution. This way, the long-term fluctuations that can be found in the laser intensity in a series of measurements are corrected. Each concentration is measured twice, as well as the solvent, the sequence solvent–solution–solvent being repeated at least three times. It is necessary to evaluate at least two different concentrations of each compound to rule out any possible mistake in the preparation of the solution. The typical process consists in the measurement of three solutions, although it is not unusual to need the measurement of several more.

(c) High voltage generator

In order to break the centrosymmetry of the considered solution, which is an isotropic liquid initially, the application of an electric field is needed. The electrical potential difference applied by a pulsed high voltage supply is at least 5–7 kV. The two electrodes are separated about 2 mm, and they are in contact with the quartz windows, and the solution. The pulsed character of the applied tension is necessary to avoid problems of hydrolysis and polarization in the solution. The pulse duration is 10 μs. The high tension pulse is synchronized with the laser pulse, whose brief duration (8 ns) compared to that of the tension allows one to treat the system as if an electric static uniform field existed in the solution. The pulse is followed by means of an oscilloscope, which delivers a reading of the field applied to the solution, as it is possible to find slightly conductive solutions or those with a very high dielectric constant where the field experienced by the solution is less than the nominal one.

2.3.1.2 Data Treatment

The experimental determination of the $\mu\beta_z$ product of a molecule is carried out in several steps. Measurements of the second harmonic intensity coming from the pure solvent are made to obtain $I^{2\omega}(0)$, alternating them with measurements of the solution at different concentrations x to obtain the coherence length $l_c(x)$ and $I^{2\omega}(x)$.

In the experiment, the Maker fringes are registered for a solution of the nonlinear molecule of interest in a certain solvent, that is, the oscillating function of $I^{2\omega}$ versus the light path, whose expression is reproduced in Equation B.15 (Appendix B):

$$I^{2\omega} = 2I_M \sin^2\left(\frac{\pi l}{2l_c}\right) \quad (2.1)$$

where $2I_M$ is the amplitude of the fringes, which depends on the concentration, and l is the length travelled by the light in the liquid, related to the movement in the X transverse direction of the cell through

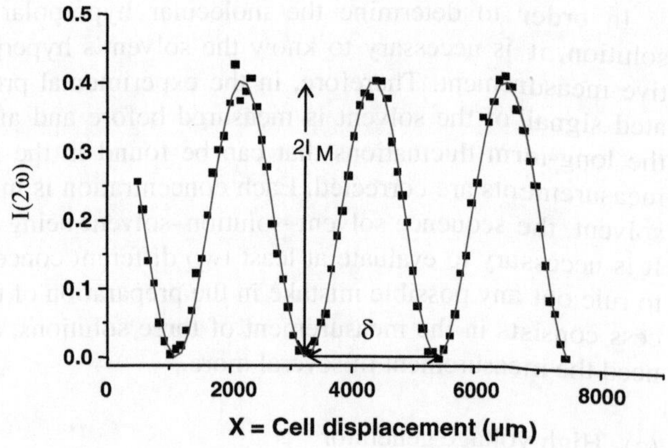

Fig. 2.10 Recording of the maker fringes obtained for a solution

$$l = 2X \tan(\alpha/2) + l_0 \tag{2.2}$$

where l_0 is the separation between the windows in the initial position (X_0) and α the angle between them.

In Fig. 2.10 a typical recording of the Maker fringes in EFISH is shown. The maximum amplitude of the fringes is identified as $2I_M$ and the spacing of the fringes as .

The curve (*C.II*) Eq. 2.1 can be fitted to the experimental data. In practice, we write it in a different way:

$$I^{2\omega} = 2A\left[1 - \cos\left(\frac{X}{B} + D\right)\right] \tag{2.3}$$

with A, B and D as fitting parameters.
A corresponds to the intensity I_M, B is related to the fringe spacing δ, which is the separation between two consecutive minima, by $\delta = 2\pi B$; and D is related to l_0, representing the initial phase.

The coherence length, l_c, corresponds to the length in the liquid in which $I^{2\omega}$ grows. In Fig. 2.10, we can see that the cell displacement must be $\delta/2 = B\pi$, thus, substituting in Eq. 2.2, we have that

$$l_c = 2\pi B \tan(\alpha/2) \tag{2.4}$$

The l_c of the solvent we use is calculated from refractive index data that can be found in the literature. With this data, the first thing to do is to fit the fringes of the solvent to calculate the angle α that the windows form in the cell. In Table B.2 (Appendix B), all the parameters used for the measurement of the $\mu\beta$ of the molecules in this thesis are gathered. Let us see now how to obtain the nonlinear coefficient from the fitting parameters.

We start from Equation *B*.16 of Appendix B:

$$\sqrt{I_M^L} = C\,[\Gamma_L l_c - K]\, E_0,$$

where C and E_0 are given in the appendix and Γ_L is the macroscopic nonlinear susceptibility.

As was mentioned at the beginning of this section, the measurements are made by alternating them with measurements on the solvent, which are used as a reference, before and after each of the concentrations that are going to be evaluated. If x is the concentration of the solution, we have

$$\sqrt{I_M^L(x)} = C(\Gamma_L(x)l_c(x) - K)E_0(x) \tag{2.5}$$

If we divide the previous equation by its equivalent for $x = 0$ (pure solvent) and find the value of the susceptibility $\Gamma_L(x)$, we find

$$\Gamma_L(x) = \frac{1}{l_c(x)}\left[(l_c(0)\Gamma_L(0) - K)\sqrt{\frac{I_M^L(x)}{I_M^L(0)}\frac{E_0(0)}{E_0(x)}} + K\right] \tag{2.6}$$

We can see that it is possible to calculate $\Gamma_L(x)$ from the fitting parameters, and from the values of K gathered in Table B.1. In order to obtain $\mu\beta_z$, we have to deduce an expression that relates the macroscopic parameter $\Gamma_L(x)$ to the hyperpolarizabilities.

To this end, we will start by characterizing the concentration of the solution, x, as mass of solute, m_S, divided by the mass of solvent, m_D.

$$x = \frac{m_S}{m_D}$$

Therefore, we can write the number of molecules per volume unit of the solute and the solvent as

$$\begin{aligned} N_D &= \frac{\rho N_A}{M_D}\frac{1}{1+x} \\ N_S &= \frac{\rho N_A}{M_S}\frac{x}{1+x} \end{aligned} \tag{2.7}$$

where N_A is Avogadro's number and ρ the density of the solution, which we assume to be equal to that of the pure solvent because of the low concentration used.

The macroscopic nonlinear susceptibility is additive with respect to the contributions of the solvent and the solute, so from expression 1.32 we can write

$$\Gamma_{ZZZZ} = Nf\gamma^0 = f(N_D\gamma_D^0 + N_S\gamma_S^0) \tag{2.8}$$

and, substituting the expressions 2.7, we obtain that for the solution and the solvent

$$\Gamma_L(x) = \frac{\rho N_A f}{1+x}\left(\frac{\gamma_D^0}{M_D} + \frac{\gamma_S^0}{M_S}x\right) \tag{2.9}$$

By setting $x = 0$ in Eq. 2.9 we obtain the microscopic hyperpolarizability of the pure solvent:

$$\gamma_D^0 = \frac{\Gamma(0)M_D}{\rho N_A f} \tag{2.10}$$

Substituting Eq. 2.10 into Eq. 2.9 and finding the value for the hyperpolarizability of the solute, we obtain

$$\gamma_S^0 = \frac{M_S}{\rho N_A f x}[(1+x)\Gamma_L(x) - \Gamma_L(0)] \tag{2.11}$$

By retrieving now expression 1.33 and finding the value $\mu_0\beta_z$, we have

$$\mu_0\beta_z = \gamma^0 5kT \tag{2.12}$$

Therefore

$$\mu_0\beta_z = \frac{M_S 5kT}{\rho N_A f x}[(1+x)\Gamma_L(x) - \Gamma_L(0)] \tag{2.13}$$

To sum up, we can say that by using expressions 2.6 and 2.13 we can determine $\mu_0\beta_z$ from the amplitude of the recording of the intensity of the second harmonic induced by the electric field.

Calculation of μβ in cases where there is absorption

In order to calculate the static hyperpolarizability value, in many cases, and nearly always when working with dipole molecules under non-absorption-conditions at the fundamental and second harmonic frequencies, the two-level model is used, as was explained in Chap. 1. This magnitude is more representative for comparison purposes, since it is independent of the frequency of measurement. The expression used to calculate $\beta(0)$ is

$$\beta(0) = \beta(2\omega)\left[1 - 4\left(\frac{\lambda_{\max}}{\lambda}\right)^2\right]\left[1 - \left(\frac{\lambda_{\max}}{\lambda}\right)^2\right] \tag{2.14}$$

where $\lambda_{\max}$ is the maximum absorption wavelength of the compound in solution, and λ the measurement wavelength (1.06 μm, 1.9 μm).

In practice, the experimental data of the concentration and the *A* and *B* parameters from the fitting of Eq. 2.3 are entered in an Excel worksheet such as the one shown in Fig. 2.11, for more efficient treatment of the data.

Calculation of $\mu\beta$ in cases where there is absorption

In the data process described so far, the possibility that the solutions absorb the light used in the excitation (ω) or generated (2ω) has not been considered. Excitation at λ = 1.9 μm avoids the absorption of the fundamental, but it is not unusual to find molecules with strong donor–acceptor substituents for which the absorption tail reaches 950 nm. Then it is very important to evaluate the absorption of the solution at 2ω, as this decreases the second harmonic intensity that we register, which leads us to underestimate the value of $\mu\beta$.

When the absorption spectrum of the molecules presents a slight absorption at the second harmonic of the measurement wavelength used in EFISH, the minima of the Maker fringes are not zero anymore, and the contrast of the oscillations decreases. The Maker fringes obtained thus were fitted to the following expression according to the model proposed by Oudar in 1977 [3]:

$$I^{2\omega} = 2OD\exp\left[-\left(\frac{\alpha_{2\omega}l}{2}\right)\right]\left[1 - \cos\left(\frac{X}{B} + D\right)\right] \tag{2.15}$$

DATE:	12/18/2009		**DAVID CELL**						
λ =	1,907	microns		**nwg=**	1,44		**G g E-14=**	2,90	
tan(alpha/2)=	0,04925802			**n2wg=**	1,45				
SOLVENT		CH_2Cl_2				**MOLECULE**		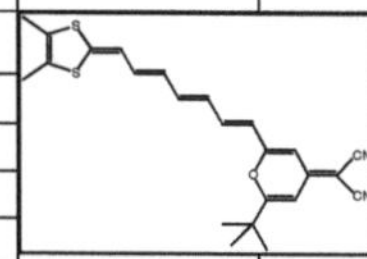	
nwl=	1,41		**Γ(0)E–14=**	8,00		**SERIES**	ELL79		
n2wl=	1,41		**lc (mic)=**	136		**MOL. W.=**	420,6		
ε(0)=	9,08		**K E-14=**	114,87		**ABS(mic)=**	0,546		
ρ=	1,3220		**f local=**	4,1973051					
FILE	x	A	B	HV (V)	lc (mic)	Γ (E–14)	gamma(E-34)	μβ (E –48)	μβ (0)(E –48)
EII7900-02	0	0,532	440,5	4,5	136,33	8			
EII7901	0,000427	1,11	436	4,5	134,94	11,29660885	97,31977021	1982,379389	1223,133186
						11,29660885	97,31977021	1982,379389	1223,133186
μβ(E–48)=		**1982**	(esus)						
FILE	x	A	B	HV (V)	lc (mic)	Γ (E–14)	gamma(E-34)	μβ (E –48)	μβ (0)(E –48)
EII7902-04	0	0,525	438,6	4,5	135,75	8			
EII7903	0,000237	0,827	436,5	4,5	135,10	9,872095073	99,55166087	2027,842444	1251,184007
						9,872095073	99,55166087	2027,842444	1251,184007
μβ(E–48)=		**2028**	(esus)						
Again c_0									
FILE	x	A	B	HV (V)	lc (mic)	Γ (E–14)	gamma(E-34)	μβ (E –48)	μβ (0)(E –48)
EII7904-06	0	0,523	438	4,5	135,56	8			
EII7905	0,0004377	1,099	435,1	4,5	134,66	11,29058074	94,77074149	1930,456311	1191,096513
						11,29058074	94,77074149	1930,456311	1191,096513
μβ(E–48)=		**1930**	(esus)						
μβ(E–48)=		***1980,2***				***μβ(0)(E–48)=***		***1237,1586***	

Fig. 2.11 Excel worksheet for the calculation of $\mu\beta$ from the values of A and B parameters measured experimentally

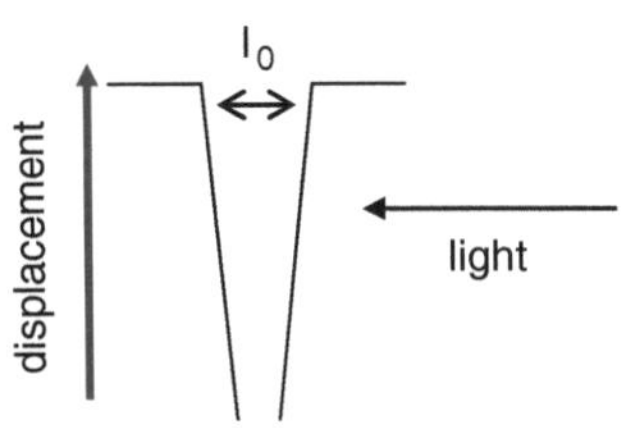

Fig. 2.12 Detail of the displacement of the quartz EFISH cell containing the solution

where $\alpha_{2\omega}$ is related to the optical density (or absorbance) as

$$\alpha_{2\omega} = \frac{OD \ln 10}{l} = \frac{A\ 2.303}{l} \tag{2.16}$$

The value of $\alpha_{2\omega}$ for each concentration used in EFISH is obtained by extrapolating from the coefficient of molar extinction, which is determined from the measurements of several more diluted concentrations prepared in an independent way.

By replacing l in the expression by its value, we obtain $l = l_0 - 2\tan(\alpha/2)X$, which would imply a cell displacement in the way shown in Fig. 2.12.

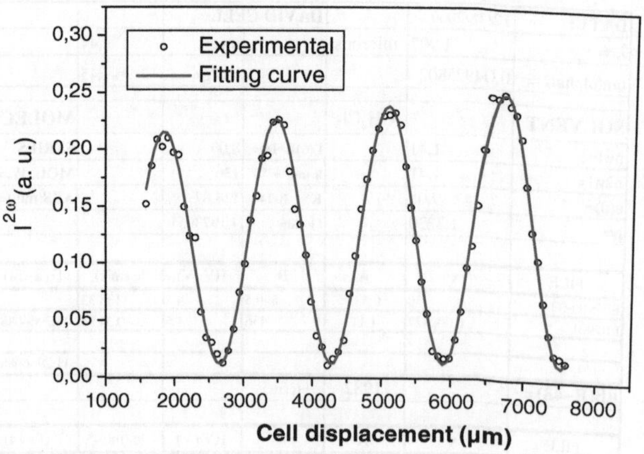

Fig. 2.13 Maker fringes fitted to Eq. 2.17. for a solution showing absorption at 954 nm

The initial point of the trajectory has to be measured, as it is the distance between the two windows at that point. Thus, we would have the following expression to adjust the fringes and to obtain A and B by fixing $G = \alpha_{2\omega}/2$ and l_0:

$$I^{2\omega} = A \exp\left[\left(l_0 - 2x \tan\frac{\alpha}{2}\right)(-G)\right]\left[1 - \cos\left(\frac{X}{B} + D\right)\right] \qquad (2.17)$$

where $2X \tan\frac{\alpha}{2} \cong X \tan\alpha \quad (\text{if } \alpha < 10°)$.

Figure 2.13 shows a recording of Maker fringes for a solution showing absorption at the frequency of the second harmonic, together with the fit to the previous equation.

2.3.2 *Thin Films: SHG in Thin Films Oriented by Corona Discharge*

The SHG measurements described below allow the nonlinear coefficients that are the second-order susceptibility tensor ($\chi^{(2)}$) components to be obtained from the Maker fringe measurements.

Two kinds of polymer systems have been studied: side-chain liquid crystals (SCLCs) with donor-π-acceptor azo chromophores in the side chain, and mixtures of azo polymers with highly efficient nonlinear chromophores.

2.3.2.1 Experimental Setup

These measurements, as well as the ones for molecular characterization, require a light source, a nonlinear spectrometer, a continuous source of high voltage (to generate the field that breaks the film's) and different sample holders (rotating, with temperature control, etc.) for each kind of measurement. These are described in more detail below.

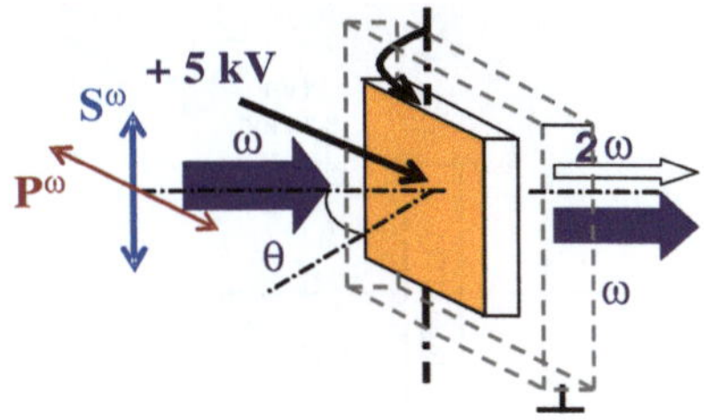

Fig. 2.14 Sample holder for the measurement of the Maker fringes in the oriented films

(a) Light source, spectrometer and cell

The light source and spectrometer used are like the ones described previously. All the measurements were made at 1.91 μm, and basically two polarizations were used, P^{ω} (horizontal in our configuration) and S^{ω} (vertical).

Regarding the sample holder, several different ones have been used, designed according to the kind of measurements required and built in the workshop of the Condensed Matter Physics Department. One of them is a fixed and thermoregulated brass holder, which is placed in a position at 40° with respect to the incident laser direction. This way, the second harmonic signal can be monitored during the orientation produced by the poling.

Another sample holder used is a rotating one, which in a routine measurement turns ± 50° around a vertical axis contained in the film, as shown in Fig. 2.14 (vertical axis). A third sample holder allows rotation around an axis perpendicular to the plane of the film and was used to measure the Maker fringes for different orientations of the film, as explained in Chap. 4.

(b) High voltage power supply. Polar orientation process by means of a field at high temperature

Thermal poling consists in orienting the sample by applying a field at high temperature and then cooling it down in the presence of the field to "freeze" the orientation. In our setup, a corona discharge is used, produced when a high voltage is applied between two conductors, at least one of which is sharp-shaped. This translates into a very intense electric field at the point, producing ionization of the surrounding air. The shape of the electrodes, distance, polarity and the gas present in the medium determine the particular features of the discharge. In our case, one of the electrodes (+) is a needle pointing towards the film. The film is placed on an ITO glass plate, which acts as the other electrode. The voltage applied to the edge of the approximately 1 mm diameter needle was between 5000 and 7000 V (see Fig. 2.15). The discharge between the needle and the film is discerned by light emission and a slight hiss. The positive ions produced seek a zone with a lower potential and reach the surface of the film (with a quite low conductivity). The distance between the electrodes (the edge of the needle and the surface of the film) is 1 cm.

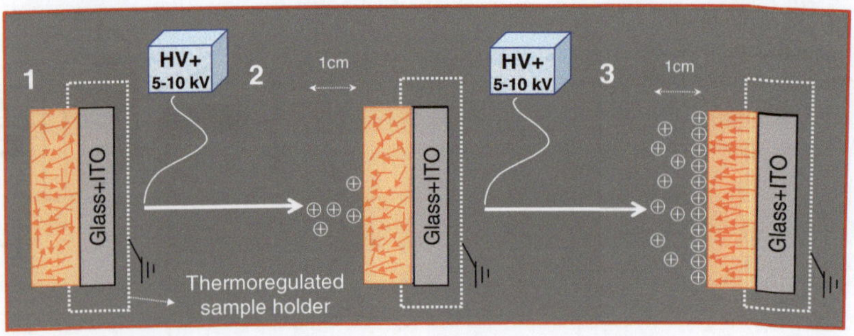

Fig. 2.15 Scheme of the orientation of the chromophores (*arrows* in the picture) by a corona discharge applied during the thermal poling

The polar molecules of a polymer film can be aligned this way, but only if the right temperature is reached so that they are sufficiently mobile ($\geq T_g$), so the process involves sample heating. This was performed in one of the holders previously described. The heating rate used was 5 °C/min, and the highest temperature reached was maintained for a period between 20 and 40 min, depending on the system studied. The cooling process was carried out with the field applied to "freeze" the orientation of the dipoles forced at high temperature [4]. A too-quick cooling process would result in the relaxation of the order achieved at high temperatures.

2.3.2.2 Data Treatment

The Maker fringe measurements are obtained with two polarizations of the excitation at frequency ω ($\lambda = 1.9\ \mu$m): P^{ω}, which is parallel to the incident plane and S^{ω}, perpendicular to it. The polarization of the generated second harmonic light will depend, in general, on the symmetry of the system and the excitation light. The Maker fringes obtained from a quartz crystal (X-cut and in the usual configuration for measurements throughout this thesis) and for an oriented film (symmetry $C_{\infty v}$) with excitation light at 1.9 μm polarized in P^{ω} have the shape shown in Fig. 2.16a and b. In both cases, the nonlinear polarization is parallel to the incidence plane, so we will call it $P^{2\omega}$. Therefore, the notation for the configurations of the measurements will be $P^{\omega}P^{2\omega}$ and $S^{\omega}P^{2\omega}$.

As can be seen, the shape of the fringes is different from that obtained in the case of a solution, where all of them had the same height (Fig. 2.10). This is because the periodic variation of the sign, which is implied by the term $\sin^2(\Psi)$ (see Chap. 1), is modified in the case of solids by a projection factor $p(\theta)$, which is the main responsible for the envelope shape (amplitude variation) of the Maker fringes in solids, that we can see in the quartz (Fig. 2.16a). The thickness to work with, in the case of films is ≈1 μm, while for quartz crystal it is ≈1 mm. Thus, in the case of quartz, we see more than one maximum, because the light covers a distance in the material several times the coherence length. However, for thin films, when turning between –50 and +50° the first maximum cannot be seen.

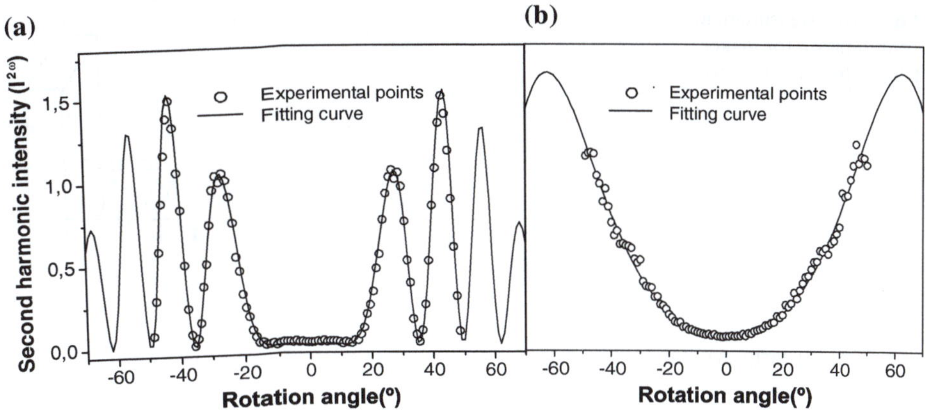

Fig. 2.16 Maker fringes for quartz glass a and for (**a**) polymeric thin film from this research (**b**)

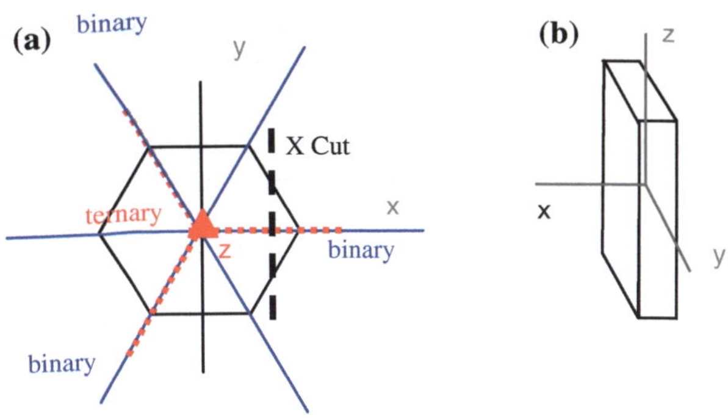

Fig. 2.17 X-cut quartz crystal. Symmetry elements

The determination of the nonlinear coefficients of the films is made from the measurements of the $I^{2\omega}$ of the sample, with respect to that generated by quartz glass. The expression that we will have to adjust to the experimental data and which allows us to find the value of the nonlinear coefficients is the one proposed by Jerphagnon and Kurtz [5] for quartz and, in general, for systems with low anisotropy values, and Herman and Hayden's one for highly birefringent films.

In this thesis, the nonlinear response of materials with different symmetry has been analysed: the X-cut quartz already mentioned (see Fig. 2.17) and polymeric thin films with uniaxial and biaxial polar orientation. Below, the features of the different experimental configurations will be detailed. Throughout, second-order nonlinear coefficients were used, d_{ijk}, whose relationship to the second-order susceptibility is given here as a reminder:

$$d_{ijk} = \chi_{ijk}/2.$$

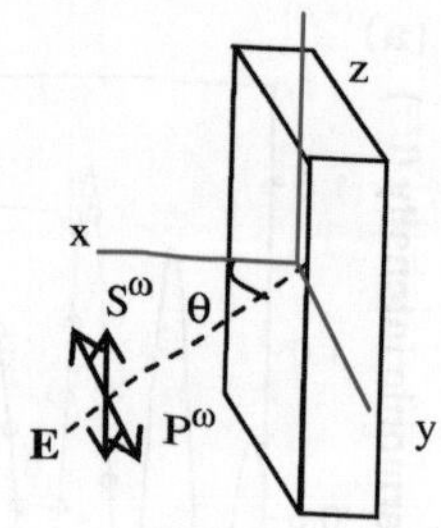

Fig. 2.18 Measurement configuration for maker fringes for a quartz crystal

(a) For a quartz crystal

Quartz belongs to the trigonal point group 32 (international notation) or D_3 (Schoenflies notation), so we will find a ternary axis (optical axis) and three binary ones (see Fig. 2.17a) as symmetry elements. X-cut crystals are cut with the large faces perpendicular to the binary *X*-axis and the turns are made around the ternary *Z*-axis, Fig. 2.17b.

For this crystal, the only nonzero nonlinear coefficients are $d_{12} = -d_{11} = d_{26}$ and $d_{25} = -d_{14}$. As $d_{14} << d_{11}$, the relationship between the polarization and the fields that generate it is written as

$$\begin{pmatrix} P_x^{(2)} \\ P_y^{(2)} \\ P_z^{(2)} \end{pmatrix} = \begin{pmatrix} d_{11} & -d_{11} & 0 & 0 & 0 & 0 \\ 0 & 0 & 0 & 0 & 0 & -d_{11} \\ 0 & 0 & 0 & 0 & 0 & 0 \end{pmatrix} \begin{pmatrix} E_x^2 \\ E_y^2 \\ E_z^2 \\ 2E_yE_z \\ 2E_xE_z \\ 2E_xE_y \end{pmatrix}$$

As the value of d_{11} is known (0.35 pm/V at 1.9 μm and 0.45 pm/V at 1.06 μm), we can use the second harmonic signal of this sample to scale the intensity obtained for samples whose NLO coefficients are unknown.

Since this material hardly shows dispersion, and its birefringence is small, we can apply the JK model (Appendix A) to fit the fringes obtained experimentally. We will use $I^{2\omega}$ measurements of the crystal in the orientation that is shown in Fig. 2.18. Therefore

$$\mathrm{I}''_{2\omega} = \left(\frac{8\pi\ \mathrm{c}}{\left((\mathrm{n}^{\omega})^2 - (\mathrm{n}^{2\omega})^2\right)^2}\right) \mathrm{d}^2\mathrm{p}^2(\theta)\,|\mathrm{E}'_{\omega}|^4\,\mathrm{t}''^4_{\omega}\mathrm{T}''_{2\omega}\sin^2\psi, \qquad (2.18)$$

where d^2 represents the nonlinear effective coefficient of the sample, whose shape depends on the precise configuration of the measurement. In this expression we know all the parameters except for the incident electric field amplitude, which will be the same for the measurement of any sample, as for the reference quartz.

Fig. 2.19 Scheme of the incidence of the excitation light on the sample

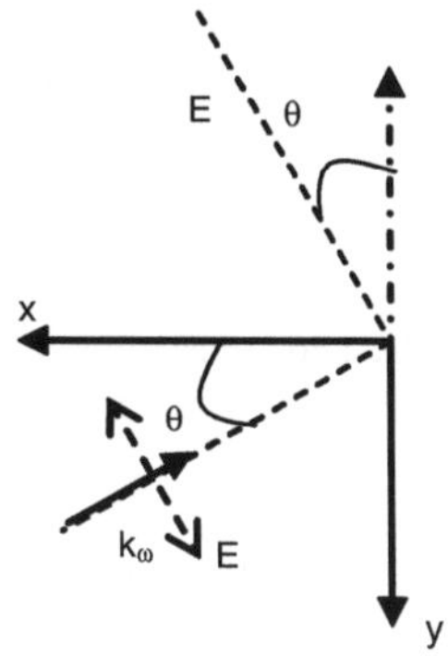

1. If the polarization of the incident light is parallel to the plane of incidence (P^ω in the picture), the components of the excitation field will be

$$E = \left(E_x, E_y, 0\right) = (E\sin\theta, -E\cos\theta, 0)$$

By substituting this in Equation A and developing the expression, we have that

$$P_x = d_{11}E^2\sin^2\theta'_\omega - d_{11}E^2\cos^2\theta'_\omega = -d_{11}E^2\cos 2\theta'_\omega,$$

$$P_y = 2d_{11}E^2\sin\theta'_\omega\cos\theta'_\omega = d_{11}E^2\sin 2\theta'_\omega,$$

$$P_z = 0.$$

We see that the $P^{2\omega}$ nonlinear polarization is contained in the *xy*-plane, the incidence plane, so we will name this configuration $P^\omega P^{2\omega}$.

As we have defined η (see Appendix A), in the case of Fig. 2.19 it would be the angle formed by the nonlinear polarization and the –*y*-axis, which would be $-(\pi/2 + 2\theta_{\omega'})$, as we infer from the expressions for P_x and P_y.

The projection factors will have then the following form:

$p_1 = 1,$

$p_2 = \cos(\theta'_\omega - \eta) = \cos(3\theta'_\omega + \pi/2) = -\sin 3\theta'_\omega$

Then: $p^2(\theta) = \sin^2 3\theta_{\omega'}$.

On the other hand, the harmonic and the fundamental wave are polarized perpendicular to the optical axis so, in the calculations, only the ordinary refractive index will be involved, both at the fundamental frequency ${n_o}^\omega$ and at the harmonic one ${n_o}^{2\omega}$.

2. If the polarization of the incident light is perpendicular to the incidence plane (S^ω in the picture), the *E* field would have just a *Z* component, that is, *E* = (0, 0, *E*). In this case, it is very quick to check, by substituting in Equation A, that the three components of the nonlinear polarization are zero, ${P_L}^{(2\omega)} = 0$, so there is no nonlinear response for this configuration.

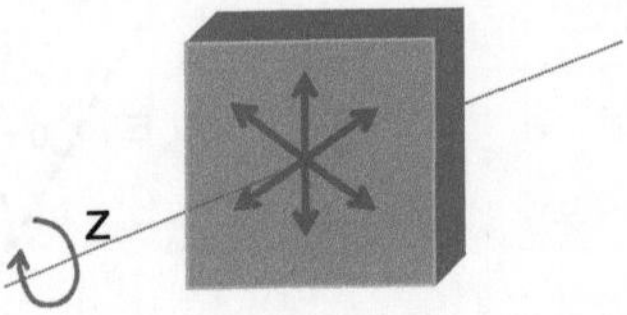

$$d_{IL} = \begin{pmatrix} 0 & 0 & 0 & 0 & d_{15} & 0 \\ 0 & 0 & 0 & d_{15} & 0 & 0 \\ d_{31} & d_{31} & d_{33} & 0 & 0 & 0 \end{pmatrix}$$

Fig. 2.20 Illustration of the symmetry a film acquires by poling orientation

(b) For an oriented film

In the case of polymeric films oriented by an electric field, two different situations have been studied. The thermal poling method, already described, applied to initially isotropic samples, leads to uniaxial systems of $C_{\infty v}$ symmetry with the axis perpendicular to the plane of the film. However, the use of photoassisted processes can vary that symmetry, giving rise, for example, to biaxial films with orientations compatible with C_{2v} symmetry.

As the medium is dispersive and the birefringence in some of the films is quite high (Δn up to 0.3), we use Equation *A*.71 (Appendix A) from the model proposed by HH for the second harmonic power at the exit of a uniaxial film, which is

$$P_{2\omega} = \frac{128\pi^3}{cA} d_{eff}^2 \left(\frac{n^{2\omega l}}{n^{2\omega}}\right)^4 \left(\frac{(n^\omega)^2-(n^{2\omega})^2}{(n^\omega)^2-(n^{2\omega l})^2}\right) \sin^2\psi \frac{[t_\omega^{a\to m}]^4 [T_{2\omega}^{m\to s}]^2 [T_{2\omega}^{s\to a}]^2}{\cos^2\gamma_{2\omega}\cos^2(\theta'_{2\omega}-\gamma_{2\omega})(n^\omega\cos\theta'_\omega - n^{2\omega}\cos\theta'_{2\omega})^2} P_\omega^2$$

where d_{eff} is the effective nonlinear coefficient, which includes the projection factor and the nonlinear coefficient, and the index expression for the bound wave is

$$n^{2\omega l}(\theta'_\omega) = n^\omega \left(\frac{1}{\frac{\sin^2\theta}{(n_e^{2\omega})^2} + \frac{(n^\omega)^2-\sin^2\theta}{(n_o^{2\omega})^2}}\right)^{1/2},$$

as explained in the Chap. 1 of this report.

In the previous expression, the effective indices and coefficients have to be distinguished for the two configurations of measurement $S^\omega P^{2\omega}$ and $P^\omega P^{2\omega}$.

(b1) Uniaxial films. Point group ∞mm (international notation), $C_{\infty v}$ (Schoenflies notation)

In the most common case of thin films oriented by a field, the C_∞ symmetry axis is the one perpendicular to its surface. In this case, the d_{IL} tensor is as given in Fig. 2.20.

By applying the Kleinman symmetry condition, we know that $d_{15} = d_{31}$, so there are only two independent coefficients, d_{31} and d_{33} (Fig. 2.21).

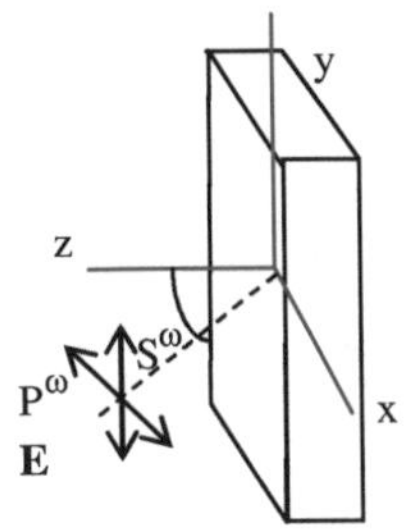

Fig. 2.21 Scheme of incidence of the excitation light on the uniaxially oriented film

$$d_{IL} = \begin{pmatrix} 0 & 0 & 0 & 0 & d_{31} & 0 \\ 0 & 0 & 0 & d_{31} & 0 & 0 \\ d_{31} & d_{31} & d_{33} & 0 & 0 & 0 \end{pmatrix}$$

We define the film axes such that the z-axis matches the optical one (normal to the film), which is perpendicular to the incidence plane. We will make the film turn around the y-axis (Fig. 2.21).

1. If the polarization of the incident light is perpendicular to the plane of incidence (XZ) S^{ω}:

$E = (0, E, 0)$,

$$\begin{pmatrix} P_x^{(2)} \\ P_y^{(2)} \\ P_y^{(2)} \end{pmatrix} = \begin{pmatrix} 0 & 0 & 0 & 0 & d_{31} & 0 \\ 0 & 0 & 0 & d_{31} & 0 & 0 \\ d_{31} & d_{31} & d_{33} & 0 & 0 & 0 \end{pmatrix} \begin{pmatrix} 0 \\ E^2 \\ 0 \\ 0 \\ 0 \\ 0 \end{pmatrix} \quad \begin{vmatrix} P_x = 0 \\ P_y = 0 \\ P_z = d_{31}E^2 \end{vmatrix}$$

The polarization is in the plane of incidence, so the measurement configuration will be $S^{\omega}P^{2\omega}$. Regarding the $p(\theta) = p_1 p_2$ factor, we have that

$$p_1 = 1,$$

since $\left|P'_{2\omega}\right| = p_1 d \left|E'_{\omega}\right|^2$, and for $P^{2\omega}$ we have

$$p_2 = \left[\cos\left((\theta'_{2\omega} - \gamma_{2\omega}) - \eta\right)\right]$$

because $\theta_{\omega}' \neq \theta_{2\omega}'$, where η, in this case, is the angle between $P^{2\omega}$ and the film surface, or x-axis in the figure, which is 90º, because it goes along the z direction. Thus

$$(p_1 p_2)^2 = \sin^2(\theta_{2\omega'} - \gamma_{2\omega}),$$

$$d_{eff}^2 = d_{31}^2 \sin^2(\theta'_{2\omega} - \gamma_{2\omega}).$$

The refractive index for the fundamental wave will be n_o^{ω} (perpendicular to the optical axis) and for the harmonic one it will be $n^{2\omega}(\theta)$, which, considering the ellipsoid of indices for the frequency 2ω (Fig. 1.5), will have the expression

$$\left(n^{2\omega}(\theta)\right)^2 = (n_o^{2\omega})^2 + [1 - (n_o^{2\omega})^2/(n_e^{2\omega})^2]\sin^2\theta.$$

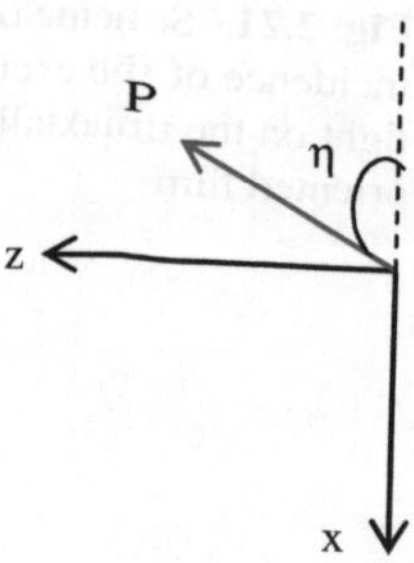

Fig. 2.22 η angle for P^{ω} excitation light in uniaxial films

The intensity generated by the film in this configuration depends only on the parameter d_{31}, which can be obtained by fitting the experimental data to Equation A.71 (the refractive indices and the thickness of the sample are known).

2. If the polarization of the incident light is parallel to the plane of incidence P^{ω}:

$$E = (-E\cos\theta,\ 0,\ E\sin\theta)\,,$$

$$\begin{pmatrix} P_x^{(2)} \\ P_y^{(2)} \\ P_z^{(2)} \end{pmatrix} = \begin{pmatrix} 0 & 0 & 0 & 0 & d_{31} & 0 \\ 0 & 0 & 0 & d_{31} & 0 & 0 \\ d_{31} & d_{31} & d_{33} & 0 & 0 & 0 \end{pmatrix} \begin{pmatrix} \mathrm{E}^2\cos^2\theta_{\omega}' \\ 0 \\ \mathrm{E}^2\sin^2\theta_{\omega}' \\ 0 \\ -2E^2\sin\theta_{\omega}'\cos\theta_{\omega}' \\ 0 \end{pmatrix},$$

$$\begin{aligned} P_x &= -2d_{31}E^2\sin\theta_{\omega}'\cos\theta_{\omega}' = -d_{31}\sin 2\theta_{\omega}', \\ P_y &= 0, \\ P_z &= d_{31}E^2\cos^2\theta_{\omega}' + d_{33}E^2\sin^2\theta_{\omega}' \end{aligned}$$

The nonlinear polarization is parallel to the plane of incidence, so we have a $P^{\omega}P^{2\omega}$ configuration. We have to consider, besides, the walk-off angle for $\theta_{\omega}{}'$ and $\theta_{2\omega}{}'$, given that the medium is birefringent and the dispersion is important (see Chap. 1):

$$\begin{aligned} (p_1 d)^2 &= \left[d_{31}\sin(2\theta_{\omega}' - \gamma_{\omega})\right]^2 + \left[d_{31}\cos^2(\theta_{\omega}' - \gamma_{\omega}) + d_{33}\sin^2(\theta_{\omega}' - \gamma_{\omega})\right]^2, \\ p_2 &= \cos((\theta_{2\omega}' - \gamma_{2\omega}) - \eta) = \cos(\theta_{2\omega}' - \gamma_{2\omega})\cos\eta + \sin(\theta_{2\omega}' - \gamma_{2\omega})\sin\eta \end{aligned}$$

where $\eta = \arctan(P_z/P_x)$ (Fig. 2.22), so we know

$$\tan\eta = P_z/P_x$$

By applying the identity:

$$\tan^2\eta + 1 = \sec^2\eta$$

We have that

$$\sin\eta = \frac{\tan\eta}{\left(1+\tan^2\eta\right)^{1/2}},$$

$$\cos\eta = \frac{1}{\left(1+\tan^2\eta\right)^{1/2}},$$

$$\sin\eta = \frac{d_{31}\cos^2(\theta'_\omega-\gamma_\omega)+d_{33}\sin^2(\theta'_\omega-\gamma_\omega)}{\left[d_{31}^2\sin^2(2\theta'_\omega-\gamma_\omega)+(d_{31}\cos^2(\theta'_\omega-\gamma_\omega)+d_{33}\sin^2(\theta'_\omega-\gamma_\omega))^2\right]^{1/2}},$$

$$\cos\eta = \frac{d_{31}\sin(2\theta'_\omega-\gamma_\omega)}{\left[d_{31}^2\sin^2(2\theta'_\omega-\gamma_\omega)+(d_{31}\cos^2(\theta'_\omega-\gamma_\omega)+d_{33}\sin^2(\theta'_\omega-\gamma_\omega))^2\right]^{1/2}}$$

By introducing $\sin\eta$ and $\cos\eta$ in p_2:

$$p_2 = \frac{\sin(\theta'_{2\omega}-\gamma_{2\omega})[d_{31}\cos^2(\theta'_\omega-\gamma_\omega)+d_{33}\sin^2(\theta'_\omega-\gamma_\omega)]+\cos(\theta'_{2\omega}-\gamma_{2\omega})d_{31}\sin(2\theta'_\omega-\gamma_\omega)}{\left[d_{31}^2\sin^2(2\theta'_\omega-\gamma_\omega)+(d_{31}\cos^2(\theta'_\omega-\gamma_\omega)+d_{33}\sin^2(\theta'_\omega-\gamma_\omega))^2\right]^{1/2}}$$

which results in,

$$\begin{aligned} d_{eff}^2 = [dp(\theta)]^2 = (dp_1)^2 p_2^2 = \sin(\theta'_{2\omega}-\gamma_{2\omega})[d_{31}\cos^2(\theta'_\omega-\gamma_\omega) \\ + d_{33}\sin^2(\theta'_\omega-\gamma_\omega)] + \cos(\theta'_{2\omega}-\gamma_{2\omega})d_{31}\sin(2\theta'_\omega-\gamma_\omega) \end{aligned}$$

Regarding the refractive indices, we have to consider the ellipsoid for the fundamental wave and for the harmonic. The index for the bound harmonic wave is defined in Appendix A as

$$\left(n^{\omega}(\theta)\right)^2 = (n_o^{\omega})^2 + [1-(n_o^{\omega})^2/(n_e^{\omega})^2]\sin^2\theta,$$
$$\left(n^{2\omega}(\theta)\right)^2 = (n_o^{2\omega})^2 + [1-(n_o^{2\omega})^2/(n_o^{2\omega})^2]\sin^2\theta.$$

In this case, the intensity of the generated second harmonic depends on d_{31} and d_{33}. By using the value of d_{31} obtained from the $S^{\omega}P^{2\omega}$ measurement, it is possible to find the value of d_{33} by fitting the measured data in $P^{\omega}P^{2\omega}$ to the expression.

(b2) Biaxial films. Point group 2 *mm* (international notation), C_{2v} (Schoenflies notation)

Irradiation with linearly polarized UV light makes the chromophores orient preferably in a perpendicular plane, so when an electric field is applied, the typical configuration of a single axis perpendicular to the film and equality in all directions in the plane of the film is changed.

For films oriented this way, there would be a *Z*-axis, perpendicular to the film and which matches the direction of the electric field orienting the dipoles, and two non-equivalent *X*- and *Y*-axes in the plane, one of which (*Y*) coincides with the polarization direction of the UV light. The chromophores would be, therefore, preferably in the *XZ* plane. In Fig. 2.23 this situation is represented in a schematic way.

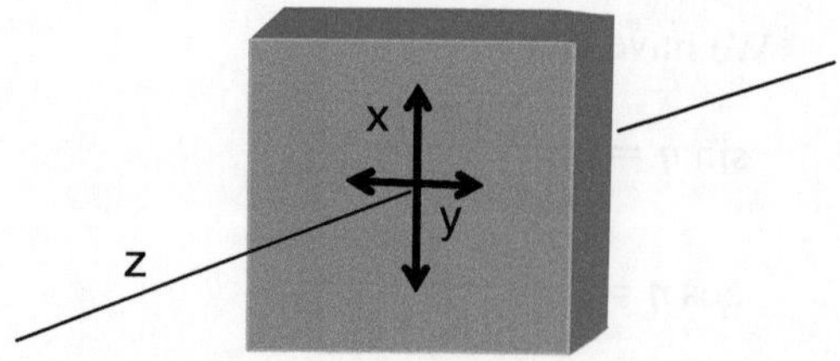

Fig. 2.23 Scheme of the axes for a film oriented by irradiation with polarized light and thermal poling

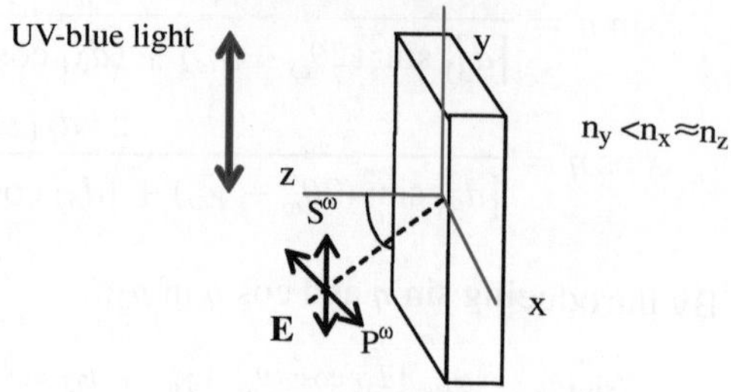

Fig. 2.24 Scheme of incidence of the excitation light in the biaxially oriented film. The irradiation direction is also marked

The components of the nonlinear susceptibility tensor would be d_{31}, d_{32}, d_{33}, d_{24} and d_{15}, and by applying Kleinman symmetry conditions ($d_{15} = d_{31}$, $d_{32} = d_{24}$), we find the tensor has the form [6]

$$d_{IL} = \begin{pmatrix} 0 & 0 & 0 & 0 & d_{31} & 0 \\ 0 & 0 & 0 & d_{24} & 0 & 0 \\ d_{31} & d_{24} & d_{33} & 0 & 0 & 0 \end{pmatrix}$$

In this case, the measurements of the Maker fringes will not be equivalent when turning around the perpendicular axis of the film (Z).

Orientation 1 (Fig. 2.24):

1. Perpendicular polarization of the incident light: S^{ω}

$$E = (0, E, 0)\,,$$

$$\begin{pmatrix} P_x^{(2)} \\ P_y^{(2)} \\ P_z^{(2)} \end{pmatrix} = \begin{pmatrix} 0 & 0 & 0 & 0 & d_{31} & 0 \\ 0 & 0 & 0 & d_{24} & 0 & 0 \\ d_{31} & d_{24} & d_{33} & 0 & 0 & 0 \end{pmatrix} \begin{pmatrix} 0 \\ E^2 \\ 0 \\ 0 \\ 0 \\ 0 \end{pmatrix} \quad \begin{array}{|l} \text{As } d_{24} = d_{32}, \\ P_x = 0 \\ P_y = 0 \\ P_z = d_{24}E^2 \end{array}$$

Therefore, we have an $S^{\omega}P^{2\omega}$ configuration. The process to obtain the projection factors is analogous to the situation for a uniaxial film, except for the fact that the coefficient that matters is d_{24}, instead of d_{31}, and that the refractive indices will be the following: For the fundamental wave, $n_y{}^{\omega}$ (perpendicular to the optical axis z), and for the harmonic,

$$\left(n^{2\omega}(\theta)\right)^2 = (n_x^{2\omega})^2 + [1 - (n_x^{2\omega})^2/(n_z^{2\omega})^2]\sin^2\theta$$

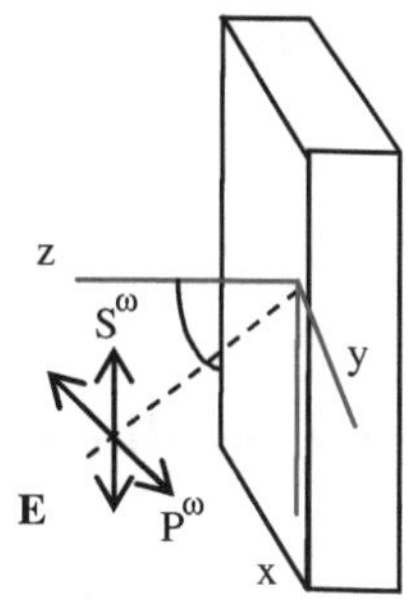

Fig. 2.25 Scheme of incidence of the excitation light in a biaxially oriented film, with the orientation perpendicular to the irradiation

2. Parallel polarization of the incident light: P^{ω}

We have the same projection factors as in the case of the $P^{\omega}P^{2\omega}$ configuration, and the same coefficients, d_{33} and d_{31}, are involved. The refractive indices are

$$\left(n^{\omega}(\theta)\right)^2 = (n_o^{\omega})^2 + [1 - (n_o^{\omega})^2/(n_e^{\omega})^2]\sin^2\theta,$$
$$\left(n^{2\omega}(\theta)\right)^2 = (n_o^{2\omega})^2 + [1 - (n_o^{2\omega})^2/(n_o^{2\omega})^2]\sin^2\theta.$$

Orientation 2 (Fig. 2.25):

1. Perpendicular polarization of the incident light: S^{ω}

$$E = (E, 0, 0),$$
$$P_x = 0,$$
$$P_y = 0,$$
$$P_z = d_{31}E^2.$$

The configuration is the same as in the uniaxial case, giving the same projection factors. The equation of the second harmonic intensity will give us the value of d_{31}, so, with this value and together with the $P^{\omega}P^{2\omega}$ measurement configuration in the other orientation, we find the value of d_{33}.

The refractive index for the fundamental wave is $n_x{}^{\omega}$ (perpendicular to the optical axis z) and for the harmonic one it is $n_x{}^{\omega}$ (parallel to the optical axis z).

2. Parallel polarization of the incident light: P^{ω}

$$E = (E, -E\cos\theta, E\sin\theta),$$
$$P_x = 0,$$
$$P_y = -2d_{24}E^2\sin\theta'_{\omega}\cos\theta'_{\omega} = -d_{24}\sin 2\theta'_{\omega},$$
$$P_z = d_{24}E^2\cos^2\theta'_{\omega} + d_{33}E^2\sin^2\theta'_{\omega}.$$

The configuration of the measurement will be $P^{\omega}P^{2\omega}$, and given the value of d_{24} calculated by using the other orientation in $S^{\omega}P^{2\omega}$ configuration, it will be possible to calculate d_{33}, and to confirm the value found before. The projection factors will be analogous to those obtained for this configuration in the other orientation.

The refractive indices are

$$\left(n^{\omega}(\theta)\right)^2 = (n_y^{\omega})^2 + [1 - (n_y^{\omega})^2/(n_z^{\omega})^2]\sin^2\theta,$$

$$\left(n^{2\omega}(\theta)\right)^2 = (n_y^{2\omega})^2 + [1 - (n_y^{2\omega})^2/(n_z^{2\omega})^2]\sin^2\theta.$$

2.4 Nonlinear Gratings

A NLO grating is a system in which the nonlinear susceptibility shows a periodic spatial modulation. The devices based on these gratings are of particular interest because they can generate second harmonic diffracted beams, which can be spatially separated from the excitation light [7]. Organic mesostructured materials with NLO properties have potential applications in photonics, as compact diode laser sources, or as waveguides [8].

In this work, the preparation of NLO gratings is based on providing a diffraction grating (previously induced by means of light irradiation) with NLO properties. First of all, we will briefly describe the formation of the gratings and then we will focus on their polar orientation.

2.4.1 Recording of Diffraction Gratings

Two monochromatic plane waves of coherent linearly polarized light are made to interfere, thus generating a periodic modulation (interference pattern) in the intensity. In the cases addressed by this thesis, linearly polarized light of 406 and 488 nm was used. If the overlapping of the two beams takes place on a photo addressable material, as in the case of polymers comprising azobenzene units, the refractive index of the film will be modified selectively, as the zones affected by light will be oriented, as we have seen before, by the so-called Weigert effect, and the dark zones will not.

In Fig. 2.26, a scheme of diffraction-grating recording is shown.

When a light beam strikes the optically recorded zone, besides the direct beam that passes directly through the sample, the diffraction maxima, which will depend on the recorded pattern, will be observed.

The intensity modulation period that is created on the plane of a thin film depends on the angle between the two beams and their wavelength (λ_g) as

$$\Lambda = \frac{\lambda_g}{\sin\theta_1 + \sin\theta_2}. \tag{2.19}$$

If we irradiate this recorded grating with a beam with wavelength λ, we will obtain, as we said before, various diffraction maxima at some specific angles. If we irradiate the grating at an angle θ_{in}, the generated diffracted maximum of order m will be appear at an angle given by

$$\sin\theta_m = \frac{m\lambda}{\Lambda} + \sin\theta_{in}. \tag{2.20}$$

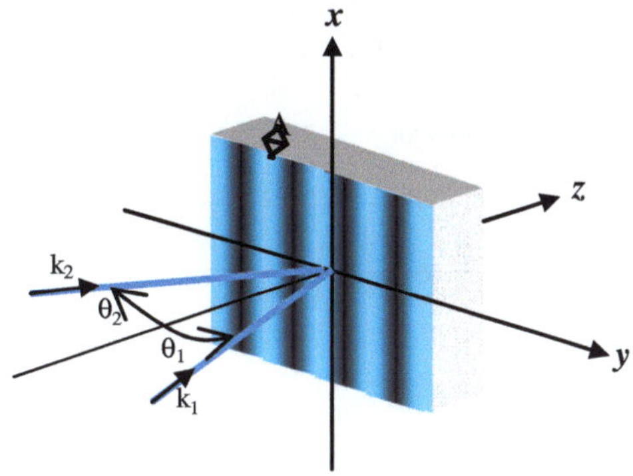

Fig. 2.26 Scheme of the interference of two beams on a film to create a diffraction grating

In order to characterize the grating, the diffraction efficiency is usually calculated as the incident light power fraction that appears in each diffraction order.

Besides the previously described recording of the gratings based on a periodic modulation of the refractive index, when an azopolymer film is irradiated with an interference pattern a modification of the surface relief takes place. This modification consists in the modulation of the film thickness with the same period as that of the interference of the two beams [9]. In this way, relief gratings with periods on the micrometre scale, and surface modulations of hundreds of nanometres can be achieved. The relief in these gratings can be increased by heating the films to temperatures in the mesophase range.

The formation of the surface relief is due to polymeric chain migrations happening on a large scale, because of the photoinduced spatial anisotropy together with a component of the electric field gradient. The presence of azobenzene side groups is a critical structural requirement for the surface distortion process. Moreover, the strong donor–acceptor structure of the main chain is also helpful, as its excited "*cis*" state shows a shorter half-life, so there are more cycles on the same time scale. On the other hand, LCs show some advantages in the recording process because they reach a higher degree of photoinduced anisotropy [10].

Among the theories developed to understand the process of formation of this surface relief, one claims that the movements in the polymeric chains are due to the pressure gradients generated when the required volume increases during azobenzene isomerization. Thus, the mass would migrate from the zones undergoing photoisomerization (irradiated zones) to those where there is less or no isomerization (dark zones). However, another theory says that the interaction between the oriented polar molecules generates an attractive force that makes molecules migrate to more ordered domains, and so an increase in relief occurs in the irradiated zones [11]. On the other hand, there are also studies in LCPs, which are similar to the ones we are going to deal with in this thesis, in which surface relief maxima in the irradiated zones were found depending on the pulsed laser power used for the recording. Thus, below a certain value of the light power the irradiated zones show maxima and above it they show depressions, assigned to polymer ablation [12].

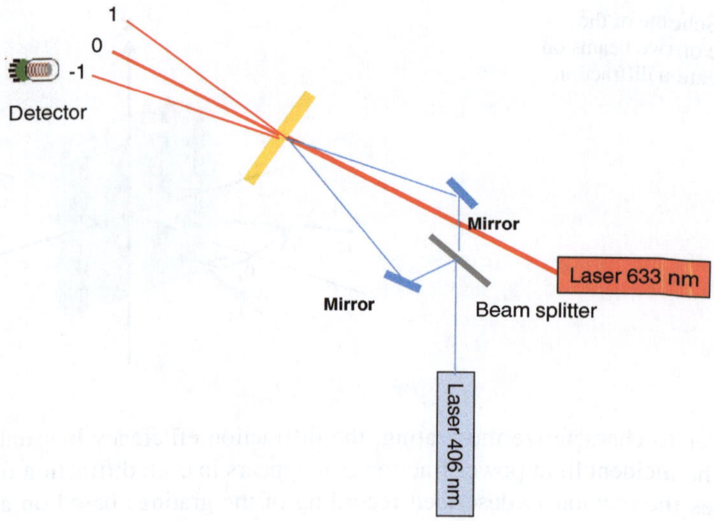

Fig. 2.27 Scheme of the experimental setup for recording nonlinear gratings

2.4.2 Recording of Nonlinear Gratings

Nonlinear relief gratings are particularly interesting, because they separate automatically the second harmonic wave from the fundamental one. On the other hand, one of the problems in achieving the maximum efficiency of the nonlinear effects in a bulk material arises from the phase difference between the fundamental wave and the harmonics, caused by dispersion. In the particular case of SHG, it is very difficult to achieve the "phase matching" condition (the efficiency of the phenomenon would be maximum then) by using only the anisotropy in the material. An alternative way to obtain large efficiencies is by introducing a periodic modulation of the susceptibility tensor in the material, with a periodicity equal to the coherence length of the waves interacting in it [13]. As we have seen before, the modulation period of the grating can be chosen by adjusting the incidence angles of the blue light used to record the grating (Eq. 2.19.).

The experimental setup used to record the relief gratings with light at 406 nm is shown in Fig. 2.27. As can be seen, the light from a diode laser at 406 nm and maximum power of 20 mW is split into two beams, which are overlapped in the film by using two mirrors, thus creating an interference pattern. In order to check whether the grating is being recorded, the recorded area is irradiated with a He–Ne beam to observe the diffraction maxima; a detector is placed in one of the first diffraction maxima with the aim of monitoring the process.

NLO properties can be induced in the grating by applying an electric field at the same time as the grating is recorded, in what would be a photoassisted poling process. Another effective method is to apply thermal poling to the grating afterwards.

The nonlinear grating efficiency is characterized by applying a horizontally polarized (P^{ω}) laser beam at 1.064 μm with an incidence angle of about 40°. The second harmonic generated light intensity at 532 nm will be detected by a Hamamatsu photomultiplier, with suitable interference filters to remove light at other wavelengths. Thus, for each film, up to three diffraction maxima intensities are measured. In order to characterize the nonlinear grating efficiency the intensity ratio of the first- and zero-order diffracted harmonic light is calculated.

References

1. R. Alicante et al., Synthesis and nonlinear optical properties of side chain liquid crystalline polymers containing azobenzene push-pull chromophores. J. Polym. Sci. Part A: Polym. Chem. **48**(1), 232 (2010)
2. R. Ulrich, R. Torge, Measurement of thin film parameters with a prism coupler. Appl. Opt. **12**(12), 2901 (1973)
3. J.L. Oudar, Optical nonlinearities of conjugated molecules. Stilbene derivatives and highly polar aromatic compounds. J. Chem. Phys. **67**(2), 446 (1977)
4. D. S. Chemla, Z. Zyss (Eds.) *Nonlinear Optical Properties of Organic Molecules and Crystals* (Academic Press, New York, 1987) 1
5. J. Jerphagnon, S.K. Kurtz, Maker fringes: a detailed comparison of theory and experiment for isotropic and uniaxial crystals. J. Appl. Phys. **41**(4), 1667 (1970)
6. F. Zernicke, J.E. Midwinter, *Applied Nonlinear Optics* (Wiley, New York, 1973)
7. L.M. Blinov et al., Polar diffraction gratings made by spatially periodic photopoling Langmuir–Blodgett films. Appl. Phys. Lett. **80**(1), 16 (2002)
8. Y. Che et al., Fabrication of surface relief grating with second-order nonlinearity using urethane-urea copolymer films. Jpn. J. Appl. Phys., Part 1 **38**(11), 6316 (1999)
9. N.K. Viswanathan et al., Surface relief structures on azo polymer films. J. Mater. Chem. **9**(9), 1941 (1999)
10. T.G. Pedersen et al., Mean-field theory of photoinduced formation of surface reliefs in side-chain azobenzene polymers. Phys. Rev. Lett. **80**(1), 89 (1998)
11. T.G. Pedersen, P.M. Johansen, Mean-field theory of photoinduced molecular reorientation in azobenzene liquid crystalline side-chain polymers. Phys. Rev. Lett. **79**(13), 2470 (1997)
12. F.J. Rodríguez, Ph.D. Dissertation, Propiedades ópticas fotoinducidas en polímeros con uidades de azobenceno Universidad de Zargoza, 2005
13. G. Martin et al., Photo-induced non-linear susceptibility patterns in electro-optic polymers. Synth. Met. **127**(1–3), 49 (2002)

Chapter 3
Nonlinear Optical Molecular Response

The work constituting this thesis includes a comprehensive characterization of the molecular hyperpolarizability by means of the EFISH technique. Part of this work has given rise to several publications in collaboration with different research groups, which designed and synthesized the molecules of the studies and also provided theoretical predictions of their NLO response. In this chapter the most interesting work from an experimental point of view, among the collection of measurements performed, is summarized. As a prior step to the study of the bulk materials, we start by analysing the properties in solution of the molecules that will become a part of the polymeric films. The study of these polymeric films will be described later in this report.

Most of the molecules studied in this thesis correspond to a basic structure in which a π-conjugated spacer acts as a bridge between an electronic density donor group and an acceptor one. Since the introduction of organic compounds to the field of nonlinear optics other kinds of structures have been explored, such as the octupolar ones, but most studies with a more practical aspect point to the dipolar nonlinear chromophores as the most suitable to be incorporated in polymeric systems with applications in electro-optic devices. In this field, mentioned should be made of the work by the group of Dalton [1], in whose laboratory some of the most efficient molecules have been synthesized and studied, with $\mu\beta$ values of the order of 10^{-43} esu measured at 1.9 μm. Although some of those molecules show intense absorption bands in the low energy zone of the visible or even near infrared spectrum, and therefore the measured $\mu\beta$ values can be affected by this proximity to a resonance situation, it is unquestionable that their incorporation in polymers has given rise to excellent nonlinear responses also at the macroscopic scale. Electro-optical modulators with very low operating voltages and a large bandwidth have been built from very efficient chromophores dispersed in polymers [1–3].

In this chapter the effects on the nonlinearity of the changes in the donor and acceptor groups and the length and nature of the spacer of the chromophores are shown.

R. Alicante, *Photoinduced Modifications of the Nonlinear Optical Response in Liquid Crystalline Azopolymers*, Springer Theses, DOI: 10.1007/978-3-642-31756-9_3,

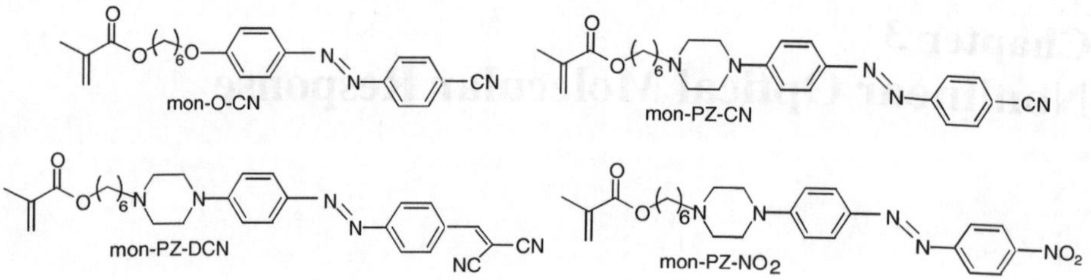

Fig. 3.1 Chemical structures of the azochromophores studied

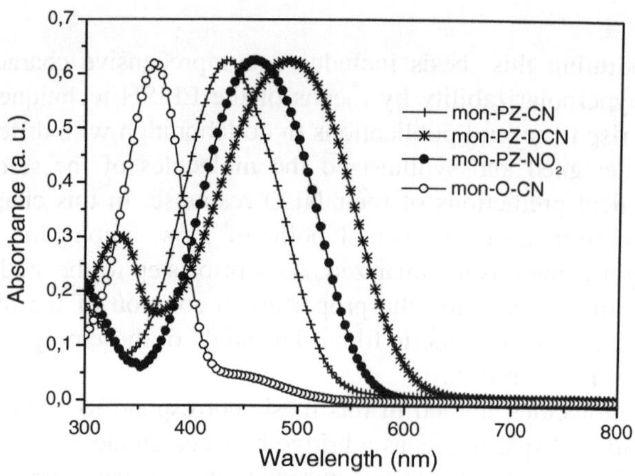

Fig. 3.2 Absorption spectra for solutions of the studied monomers in CH_2Cl_2

3.1 Study of NLO Chromophores with Azo Groups

Considering that one of the issues addressed in this thesis is the relationship between the molecular properties and the nonlinear response of the bulk material, it is necessary to determine the molecular hyperpolarizability values of the monomers used in the synthesis of the polymers that will be studied in Chap. 4. We begin by studying the properties of several compounds bearing azobenzene in the π-conjugated bridge. The molecular structures that we are going to deal with can be seen in Fig. 3.1.

We work with these structures in order to obtain photoaddressable systems for nonlinear optics, in which the response is determined by the introduction of gradually stronger acceptor groups, such as the cyano, nitro and dicyano ones. All of them, except the one containing the dicyano group, show smectic phases.

The corresponding absorption spectra of the monomers (see Fig. 3.2) show a clear bathochromic shift when the strength of the acceptor group in the compound is increased. Besides, for mon-O-CN, the donor group is also the weakest

Table 3.1 Absorption-maximum wavelengths and nonlinear hyperpolarizabilities, in CH_2Cl_2, for the monomers studied [4]

Compound	λ_{max} [nm])	$\mu\beta[10^{-48}esu]^a$	$\mu\beta$ (0) $[10^{-48}esu]$	$\mu\beta[10^{-48}esu]^b$	$\mu\beta(0)$ $[10^{-48}esu]$
mon-O-CN	365	180	80	90	70
mon-PZ-CN	430	950	280	340	260
mon-PZ-NO_2	460	–[c]	–[c]	440	320
mon-PZ-DCN	490	–[c]	–[c]	750	510

Experimental error estimated for $\mu\beta$ ±15 %. [a]Measurements at 1.06 μm. [b]Measurements at 1.91 μm. [c]The bathochromic shift in the absorption spectrum makes it impossible to obtain reliable values at 1.06 μm

one, so it shows the absorption maximum in the higher energy region. The wavelength values at which the absorption is maximum in solutions of the monomers in CH_2Cl_2 are gathered in Table 3.1.

In Table 3.1, the molecular hyperpolarizability values for the monomeric units obtained by means of the EFISH technique can also be found. Where possible, two wavelengths were used, 1.06 and 1.91 μm, to analyse dispersion effects.

In order to calculate the $\mu\beta$ values of mon-O-CN and mon-PZ-CN at 1.06 μm, the absorption at 532 nm of both compounds in the solvent used for the EFISH measurements had to be taken into account. To this end, at least five different concentrations were measured, and the slope of the fitting line of the absorbance values (at 532 nm) versus concentration was calculated. The values for the slopes were 192 and 1,770 (l/cm mol) for mon-O-CN and mon-PZ-CN, respectively. If it is assumed that the linearity between the absorption and the concentration is maintained for the values of the EFISH solutions, it is possible to determine the absorption coefficient $\alpha_{2\omega}$ needed to fit the experimental fringes obtained for each solution of mon-O-CN and mon-PZ-CN, as explained in Chap. 2 of this thesis.

The static hyperpolarizability values $\mu\beta(0)$ were deduced by using the two-level dispersion model (Eq. 1.36). As can be observed in Table 3.1, the results arising from experimental values obtained at different wavelengths were very similar, which supports the validity of the two-level model for these kinds of dipolar compounds with rod-like structure.

As expected, the introduction of the piperazine group as a donor, causes an increase in the NLO response, which also increases with the strength of the acceptor moiety (DCN > NO_2 > CN). These results are consistent with reported studies for a series of donor-π-acceptor azochromophores containing dialkylamine as the donor group and the acceptor moieties we have included, among others [5]. These monomers will be polymerized in order to study their macroscopic NLO properties.

In addition, the study of certain azochromophores that were designed as V-shaped "bent-core" liquid crystals was proposed [6]. In principle, this geometry is suitable for maximizing the macroscopic response [7, 8], as it promotes the formation of polar smectic layers, leading to non-centrosymmetric phases.

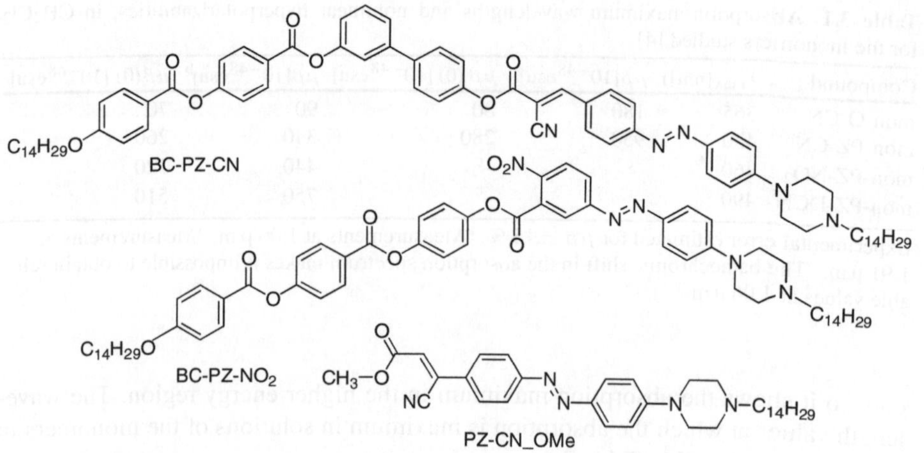

Fig. 3.3 Chemical structure of the 2D azocompounds

Table 3.2 Absorption-maximum wavelength and hyperpolarizabilities for the 2D azocompounds and PZ-CN_OMe

Compound	λ_{max} [nm]	$\mu\beta[10^{-48}$esu]	$\mu\beta$ (0) [10^{-48}esu]
BC-PZ-CN	485	560	390
BC-PZ-NO_2	470	630	440
PZ-CN_OMe	472	470	330

The synthesized compounds are azochromophores bearing donor and acceptor groups similar to the ones studied previously, but with different structures. As can be observed in Fig. 3.3, BC-PZ-CN and BC-PZ-CNO_2 have 2D structures, in which the donor-π-acceptor system incorporated in the side structure is quite similar to the 1D molecules studied previously, mon-PZ-CN and mon-PZ-NO_2, respectively. BC-PZ-CN shows a columnar mesophase, while in the case of BC-PZ-NO_2, liquid crystalline phases were not observed.

In this case, the EFISH measurements were carried out at 1.9 nm with dichloromethane as solvent, and the results are shown in Table 3.2. The incorporation of the piperazine donor group and the cyano and nitro acceptors gives rise to $\mu\beta(0)$ values an order of magnitude larger than the ones usually measured in V-shaped liquid crystals [9]. In this table, the increase in the response on going from the most linear molecular structure PZ-CN-O-Me to the banana-shaped system can be seen. On the other hand, if we compare the $\mu\beta$ value of PZ-CN-O-Me with that obtained for mon-PZ-CN (in the previous section), a slight increase in the response is observed, which is related to the displacement of the absorption band maximum to longer wavelengths.

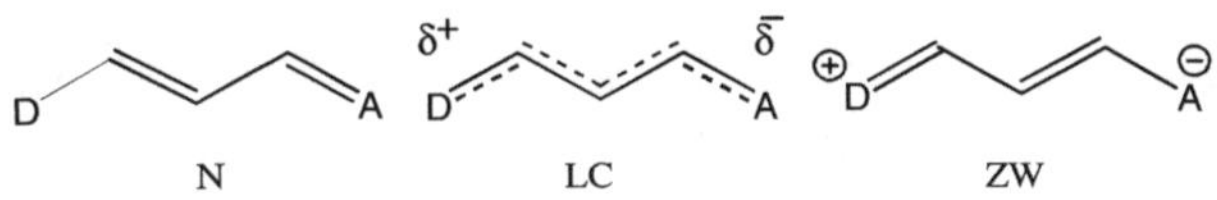

Fig. 3.4 Representation of the model resonance forms for the merocyanine-type structures

3.2 Characterization of Highly Efficient Chromophores

Throughout this thesis a wide variety of molecules have been characterized, in the framework of the collaboration with the group of Prof. Javier Garin from the Organic Chemistry Department of the University of Zaragoza. The experimental results from EFISH for several groups of molecules, in which the influence of the modification of the donor and acceptor groups, the nature and length of the conjugated chain as well as the solvent is reflected in the $\mu\beta$ values, will be presented below. Most of the studied compounds are referred to as merocyanines, which are defined as a polyene with donor and acceptor sustituents and an odd number of π centres, so that $n+3\pi$-electrons are delocalized in the system.

As was explained in Chap. 1, in many donor-π-acceptor systems the charge transference controls the NLO response. As a first approximation, this allows the value of the hyperpolarizability to be discussed in terms of a scheme in which only the fundamental and the first excited state take part. As we explained previously, these states can be described as combinations of the extreme resonance structures represented in Fig. 3.4.

In the design of the molecules for nonlinear optics, the control of the fundamental state polarization as a function of the relative contribution of the two extreme forms is considered a fundamental issue. The structural parameter BLA (bond length alternation) quantifies this contribution, by characterizing the degree of delocalization on the π-system. In the 1990s, Marder and co-workers established a relationship between BLA and the sign and magnitude of the molecular hyperpolarizability, as shown in Fig. 3.5. For instance, the fundamental state of a polyene with very weak D and A substituents will be dominated by the neutral form, with a large value of BLA. This system presents a positive β but small in absolute value. As the donor–acceptor strength increases, the charge separation resonance form grows in importance, providing smaller BLAs, until both resonance forms contribute equally and BLA becomes 0. In the case of the latter, the hyperpolarizability value also becomes zero, analogously to the case of a symmetric molecule. If we keep increasing the polarization of the fundamental state the transference charge resonance form will control the fundamental state structure, giving rise to negative BLA values. In this way, we will have RHS and LHS compounds, which means that they are placed on the right-hand side (negative β values) or left-hand side (negative β values), respectively, of Marder's curve. As can be seen in Fig. 3.5, different regions of the curve are distinguished and the compounds can be related to them.

Among the measured compounds we find merocyanines that, inside the same family, can be placed in different regions of Marder's curve, depending on the substituents or the chain length. Systems have also been explored that are located near

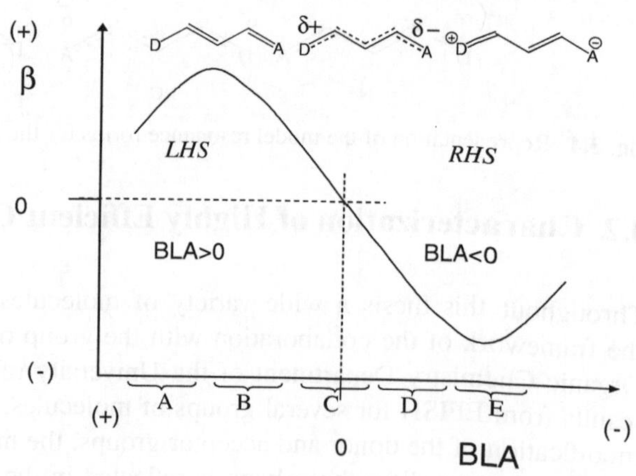

Fig. 3.5 Marder's curve, which relates the hyperpolarizability to BLA

the cyanine limit (CL) and that pass from one side to the other of this intermediate situation depending on the solvent, swapping between positive and negative values when the solvent is changed. Apart from the essentially one-dimensional (1D) D-π-A compounds, D-π-A-π-D type V-shaped (2D) molecules with the acceptor in the V vertex have been studied.

Finally, from an experimental point of view, it is necessary to note that all the values of $\mu\beta$ presented in this chapter, have been measured under conditions in which various measurements of dilute solutions of Disperse Red 1 (DR1) (reference molecule) converge on $\mu\beta(0) = 450 \times 10^{-8}$ esu [10] in CH_2Cl_2. In addition, the estimated error inherent in the experiment, unless otherwise specified, is ± 15 %.

Below, we will proceed to detail the results obtained, where the compounds are grouped according to the following scheme:

1. Merocyanines with dicyanothiazole as a proaromatic acceptor Influence of the different donors, chain length and nature.
2. Merocyanines with 1,3-dithiol-2-ylidene as a proaromatic donor Influence of the different acceptors, chain length and nature.
3. Merocyanines with pyran derivatives as proaromatic donors Influence of the different position of the group, different acceptors, chain length and acceptor group's substitution.
4. Merocyanines with isophorone, pyran and dihydropyran as spacers.
5. Merocyanines with thiophene spacers Influence of the chain lengthening on both sides of the thiophene.

3.2.1 *Merocyanines with Dicyanothiazole as a Proaromatic Acceptor*

Much effort has been devoted to the search for molecules with high and positive β values, that is, that are positioned on the left-hand side of Marder's curve

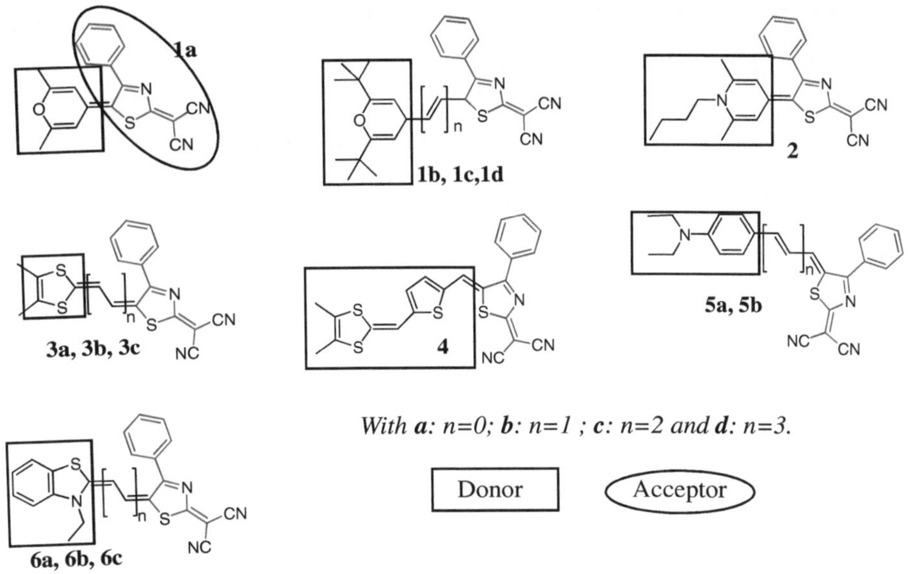

Fig. 3.6 Chemical structures of the compounds with the proaromatic acceptor dicyanothiazole

(LHS). The systems having a ground state with a significant zwitterionic contribution have been much less studied. However, the right-hand side of the curve offers many possibilities since strongly polar molecules are found there, which can give rise to very high values of β although negative.

In order to investigate this line, the molecules presented below were synthesized. Their common feature is the presence of a quinoid thiazole and a proaromatic donor. It was therefore proposed to characterize, by means of the EFISH technique, the merocyanines whose structure is shown in Fig. 3.6 with a quinoid thiazole linked to the dicyano acceptor group and different donors, which are also proaromatic [11, 12].

As mentioned, the compounds differ in the electron donor group and the spacer. Molecules **1a-d** present 4*H*-pyran-4-ylidene, and **3a–c** and **4** a 1,3-dithiol-2-ylidene. Compound **2** was synthesized for the sake of comparison between two donors: piranylidene and pyridilidene. In addition, compound **4** allows the effect of the introduction of a thiophene ring in the spacer to be analysed. The substitution of aromatic spacers for linear ones in these kinds of molecules improves the limited thermal and photochemical stability of the former. Compounds **5a** and **5b** are included to study the influence of an aromatic donor. Finally, compounds **6a–c** with a benzothiazole as a proaromatic donor are included.

In Table 3.3 the values of the wavelengths corresponding to absorption maxima in two different solvents are summarized. Moreover, in Table 3.4, we show the results for $\mu\beta$ at 1.9 μm in dichloromethane. The values of $\mu\beta(0)$ are obtained using the two-level model, as explained in Chap. 2.

Table 3.3 Absorption-maximum wavelength of the compounds with a dicyanothiazole proaromatic acceptor

	λ_{max} (CH_2Cl_2)	λ_{max} (DMSO)		λ_{max} (CH_2Cl_2)	λ_{max} (DMSO)
1a	550,583	564	**3c**	734,803	795
1b	634,683	666	**4**	725(s),795	822
1c	726,790	756	**5a**	598(s),638	657
1d	821,900	–[a]	**5b**	679,732	754
2	533	494	**6a**	592(s),636	608
3a	561(s),596	597	**6b**	674(s),739	669
3b	648,699	695	**6c**	850	681

[a] unstable, it decomposes; *s* shoulder

Table 3.4 Hyperpolarizabilities measured in CH_2Cl_2 and DMSO for the compounds with dicyanothiazole proaromatic acceptor

	$\mu\beta[10^{-48}$esu]	$\mu\beta$ (0) $[10^{-48}$esu]		$\mu\beta[10^{-48}$esu]	$\mu\beta$ (0) $[10^{-48}$esu]
1a	−360	−200	**3c**	1900	460
1b	−1070 (−1200)	−450 (−540)	**4**	5440	1370
1c	−3900 (−5960)	−1010 (−1860)	**5a**	1610	790
1d	−38000	−3200	**5b**	5940 (3040)	2080 (960)
2	−1150	−730	**6a**	(−820)	(−440)
3a	−210	−115	**6b**	(−4300)	(−1900)
3b	300 (−900)	120 (−370)	**6c**	(−5300)	(−2300)

DMSO measurements: Values in parentheses ()

Considering the absorption spectrum in both solvents, all compounds show a band of intramolecular charge transfer with λ_{max} in the visible, but with wide bands that extend to the NIR in some cases. In addition, many of the bands are composite or structured, showing shoulders and even different bands at high frequencies. The absorption maxima increase significantly in all series when the polyene chain is lengthened. In compounds **1**, **3** and **6** the increase on adding a double bond is $\geq$100 nm. Compounds **1**, **2** and **6** show a clear negative solvatochromism on going from dichloromethane to dimethyl sulfoxide (DMSO), while **3** does not show a clear trend, which is compatible with fundamental states being close to the CL.

Thus, one can predict that compounds **1**, **2** and **6** will be on the right-hand side of Marder's curve (RHS), and so they will show negative hyperpolarizability values. The experimental results are compiled in Table 3.4.

Compounds **1**, **2** and **6**, in effect, resulted in negative $\mu\beta$ values. In the 1 and 6 series, it can be seen how the absolute value of $\mu\beta$ increases with the lengthening of the conjugated chain. If **1a** and **2** are compared, the absolute values of $\mu\beta$ indicate that the strength of the donor decreases on going from pyridilidene to piranylidene, so compound **2** would be located further to the right in Marder's curve, that is, in the C/D region. The larger values measured in DMSO for **1b** and **1c** than for **1a** confirm that the compounds would be placed in the C region in dichloromethane, approaching the D region when the solvent polarity increases, as in the case of DMSO, for which higher absolute values are obtained. It can be said as well by looking at the compounds **1b** and **6b** that benzothiazole is also a better donor than piranylidene. Finally, since it was not possible to measure **6** compounds in CH_2Cl_2 owing to their low solubility, it is not possible to assign them to a specific area of the curve, but would be in regions D and E.

On the other hand, compounds **5**, with an aromatic donor, are predominantly neutral and are placed on the LHS in Marder's curve. Proof of this is the smallest value of $\mu\beta$ obtained in DMSO, since the chromophore approaches the CL, thus lying in region B of the curve. The incorporation of a thiophene cycle in the conjugated chain in compound **4** significantly improves the NLO properties, with respect to the analogous compound with a linear chain, **3c**, placing the compound in region B of the curve.

As for the compounds of series **3**, the results in CH_2Cl_2 show a change in the sign of $\mu\beta$ on going from **3a** to **3b**, indicating that the CL is exceeded from RHS. The larger and positive $\mu\beta$ value for **3c** would confirm the leap to LHS. In addition, the value for **3b** in DMSO becomes negative, which means that these compounds are located in the B/C region of the curve. This corresponds to compounds with ground states near the CL, as was mentioned in the discussion of solvatochromism.

3.2.2 *Merocyanines with 1,3-Dithiol-2-ylidene as a Proaromatic Donor*

In this section, the study of another series of compounds, whose common feature is that they have 1,3-dithiol-2-ylidene as a donor group, such as compounds **3** and **4**, is described in detail. However, they incorporate different acceptors, which are numbered below: tricyanofuran (TCF), phenyloxazolones, isophorones (with dicyano and thiobarbituric), and pyrans (with dicyano and thiobarbituric).

The compounds have been grouped to show the results such that in each group the acceptor is the same, and their substituents and the length and nature of the conjugated chain can change.

3.2.2.1 With Tricyanofuran as the Acceptor Group

In Fig. 3.7, the structures of the compounds bearing dithiol as the donor group and TCF as the acceptor group are shown [13]. Also, in Table 3.5 the results of the absorption-maximum wavelength of the mentioned compounds, as well as the $\mu\beta$ values obtained by EFISH at 1.9 μm, are shown.

From the preliminary analysis of the data regarding the **7** and **8** series, the conclusion is that the lengthening of the conjugated chain leads to bathochromic shifts and better NLO properties of the molecule. Likewise, though the absorption of the compounds **8** is displaced to lower energies, no improvement in the properties can be seen when the methylenic substituents of TCF are replaced by trifluoromethylenes (TCF-CF_3), the former being a stronger acceptor. The compounds **7** show

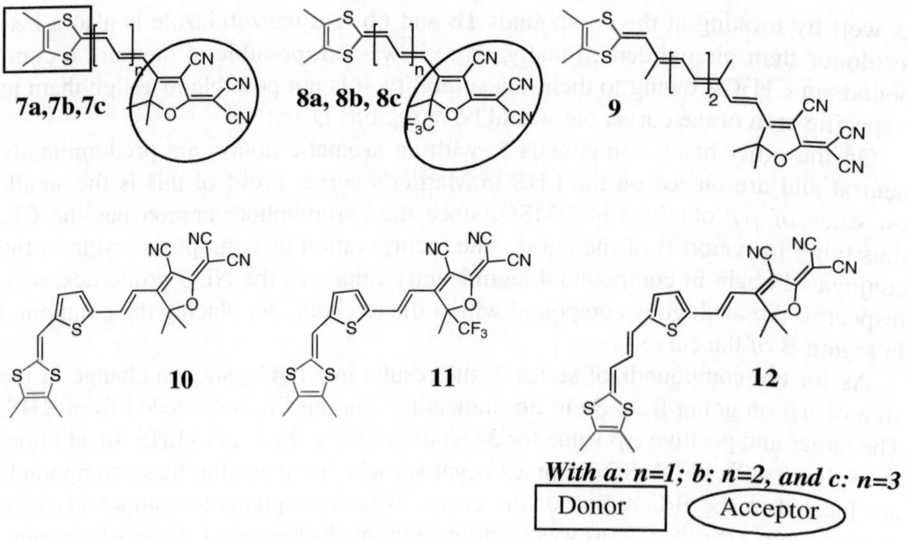

Fig. 3.7 Chemical structure of the compounds with dithiol donor group and TCF acceptor

Table 3.5 Absorption-maximum wavelengths and hyperpolarizabilities, in CH_2Cl_2, of the compounds with dithiol donor group and TCF acceptor

	λ_{max} (CH_2Cl_2)	λ_{max} (DMSO)	$\mu\beta[10^{-48}$esu]	$\mu\beta$ (0) $[10^{-48}$esu]
7a	600(s),653	664	1300	610
7b	678,732	751	5560	1940
7c	721,835(s)	710,869	16820	6170
8a	630(s),688	688	910	380
8b	711(s),791	789	5270	1360
8c	789(s),894	893	25800	2430
9	742,968(s)	–	31000	10370
10	723	711	5440	1980
11	793	815	11200	2850
12	716	691	5600	2100

positive solvatochromism, while for the **8** ones this is almost zero. This points to a strong polarization of the fundamental state and a proximity to the CL, reaching the B/C region of the curve, so the response is lower than in the compounds **7**. As the measurement of $\mu\beta$ in a more polar solvent can help place these compounds in one region or another, we tried to measure **8b** in DMSO. However, the compound decomposed, so dimethylformamide (DMF) was used, which is a more polar solvent than dichloromethane. The result was $\mu\beta = -800 \times 10^{-48}$ esu, which shows that the change of the solvent makes the compound exceed the CL and be defined as RHS. Measurements of the analogous non-fluorinated compound **7b** were made in DMF and DMSO, showing such a decrease that $\mu\beta = 1{,}500 \times 10^{-48}$ esu in the first one and $\mu\beta = 2{,}900 \times 10^{-48}$ esu in the second. Therefore, these compounds are LHS, and would be placed in region B zone of Marder's curve. However, similar compounds with π-TCF acceptor, but with a different proaromatic donor, pyrid-4-ylidene, have been categorized in the literature as RHS [14]. On the other hand, compound **9** showed the largest $\mu\beta$ value known for chromophores based on 1,3-dithiol [15, 16].

It is interesting to note that when the thiophene group is included in the conjugated chain, $\mu\beta$ clearly decreases, as can be seen on going from compound **7c** to **10** and from **8c** to **11**. It is also possible to see that there is barely any bathochromic shift in the absorption in either case.

Finally, a comparison can be made between compounds discussed in Sects. 3.2.1. and 3.2.2.1. It is possible to see that **4** and **10**, on the one hand, and **3** and **7**, on the other, differ only in the acceptor group: tricyanopyran and dicyanothiazole. The values of $\mu\beta$ for **4** and **10** are very similar, however $\mu\beta(0)$ is higher for **10**, as its absorption,λ_{max}, is shifted to the blue. Both compounds are LHS, but **4** is probably placed in region B/C and **10** in region A/B. In the other pair, the compounds **3** are close to the CL, as has been previously mentioned, and compounds **7** show a larger $\mu\beta$. Because of all this, we can say that TCF acts as a stronger electron acceptor than dicyanothiazole.

3.2.2.2 With Phenyloxazolone as the Acceptor Group

The following study comprises compounds with dithiol donor and phenyloxazolone acceptor with different chain length (Fig. 3.8). Besides, the impact of cycling the spacer by using an isophorone will be analysed [17].

The obtained results are shown in Table 3.6; they will be commented on below.

First of all, we will pay attention to the trends observed in Table 3.6, and subsequently, the efficiency of the phenyloxazolone acceptor will be analysed in comparison to other acceptors.

All the compounds show a positive solvatochromism on passing from CH_2Cl_2 to DMSO, which agrees with the positive results of $\mu\beta$, characteristic of LHS molecules. In addition, compounds **13** show bathochromic shifts when the chain length increases. The trend of increasing $\mu\beta$ with chain lengthening is fulfilled for both compound series **13** and **14**.

The effect of including isophorone can be studied by comparing compounds with the same conjugated chain length, in this case **13b** and **14d** and also **13c** and **14a**. On passing from a linear chain to a cyclic one, the λ_{max} displacement is almost zero or negative in the first case, so no clear trend can be determined. In fact, for the first case, $\mu\beta$ decreases and for the second one, it increases.

3.2.3 Merocyanines with Pyran Derivatives as Proaromatic Donors

The compounds in this study will be grouped such that in the first block the pyran derivative acts as an electron donor, in the second as a spacer, and in the third as both simultaneously.

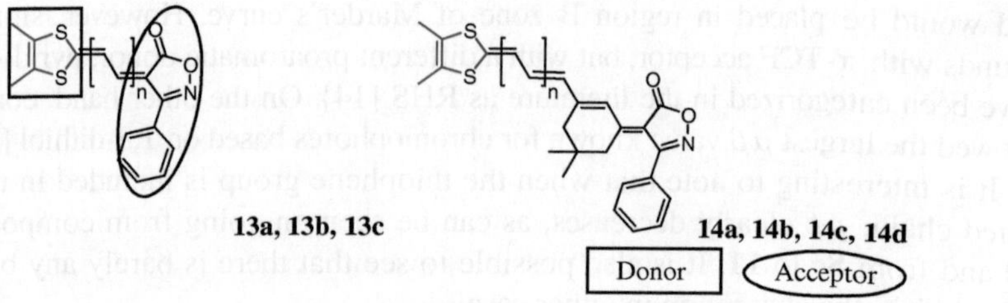

Fig. 3.8 Chemical structure of the compounds with dithiol donor and phenyloxazolone acceptor

Table 3.6 Absorption-maximum wavelengths and hyperpolarizabilities, measured in CH_2Cl_2, of the compounds with dithiol donor group and phenyloxazolone acceptor group

	λ_{max} (CH_2Cl_2)	λ_{max} (DMSO)	$\mu\beta[10^{-48}$esu]	$\mu\beta$ (0) [10^{-48}esu]
13a	547	520(s),553	250	150
13b	587,621(s)	598(s),637	1100	565
13c	633	646,726(s)	2200	1090
14d	622	632	720	380
14a	582(s),609	590(s),622	2750	1480
14b	606	624	4250	2230
14c	614	626	5540	2840

3.2.3.1 As an Electron Donor Group

In the following group of compounds, structures containing 4*H*-pyran-4-ylidene as a proaromatic donor group and different acceptor groups and conjugated chain lengths are presented [18], as can be seen in Fig. 3.9.

The maximum absorption and $\mu\beta$ results in dichloromethane solution are presented Table 3.7. In this table the absorption maxima show a shift to the red as the polyenic chain is enlarged, as we have seen so far in the other families. This bathochromic displacement is accompanied by an increase of the NLO response in all cases, except on passing from **15a** to **15b**. The solvatochromism is, generally, very small. For **17** and **19** it is very small and positive, however, in **18** it is relatively large and negative. In series **15** and series **16**, it is possible to see that for the short ones, solvatochromism is small and negative, but, when the chain is enlarged, it is positive, as the cyanine character decreases when the conjugation length is increased.

The longest structures give positive and high $\mu\beta$ values, which places them in region B of Marder's curve. On the other hand, the shortest chains keep an important zwitterionic contribution, which places them closer to or across the CL in region C. The wavelength of the maximum decreases with the acceptor strength in the following way approximately: **19** > **17** > **18** ≅ **15** ≅ **16**.

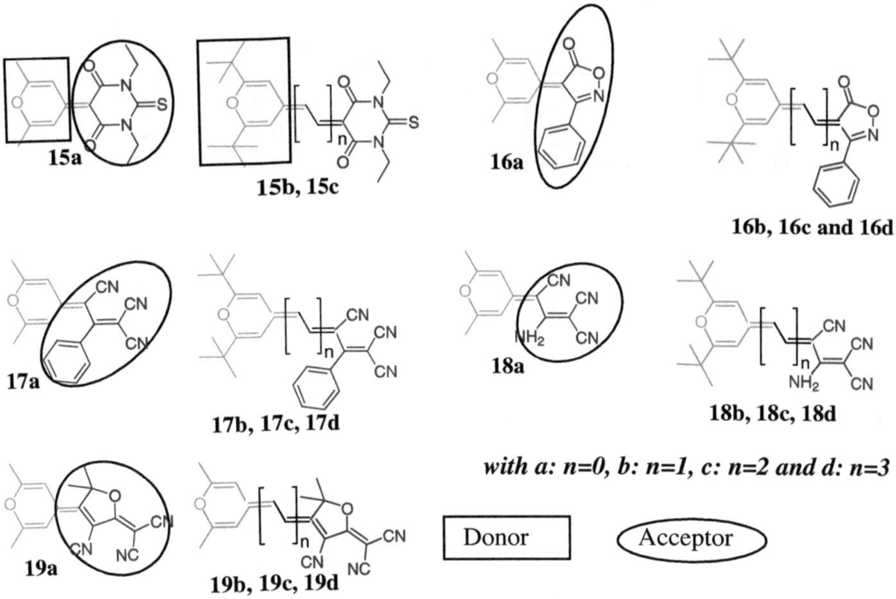

Fig. 3.9 Chemical structures of the compounds with donor groups derived from the pyran group

Table 3.7 Absorption-maximum wavelengths and hyperpolarizabilities, measured in CH_2Cl_2, of the compounds with pyran donor group

	λ_{max} (CH_2Cl_2)	λ_{max} (DMSO)	$\mu\beta[10^{-48}$esu]	$\mu\beta$ (0) $[10^{-48}$esu]
15a	392(s), 409	407	−50	−40
15b	501,530	502, 530	−90	−60
15c	584,628	585, 627	600	30
16a	397	396	–[a]	–
16b	498,530	502, 529	100	64
16c	579,623	583, 626	740	380
16d	610(s),654,715	663, 725	3500	1320
17a	455	460	60	40
17b	535,568,601(s)	536(s), 570, 600(s)	210	120
17c	629,678	629(s), 678	1550	670
17d	719,787	718(s), 782	7500	2000
18a	382	358	25	20
18b	506,532	485	190	120
18c	584,625	552, 603(s)	705	360
18d	643,719	558, 713(s)	2650	980
19a	513	517	140	90
19b	592,639	595, 637	630	310
19c	675,740	680, 739	4500	1520
19d	688(s),753,841	763, 847	17400	3110

[a] The harmonic signal intensity of the solution can not be distinguished from the solvent's one

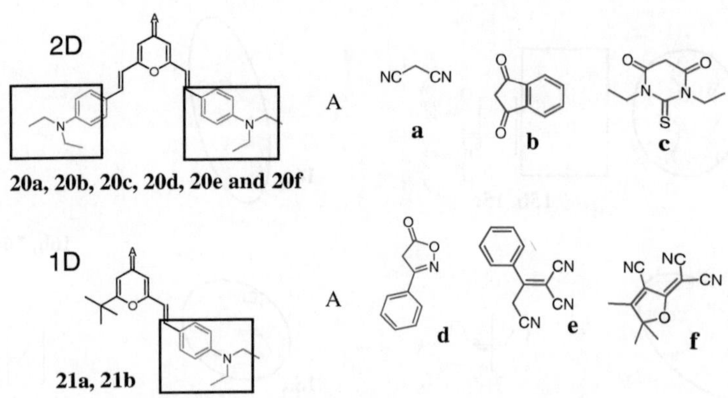

Fig. 3.10 Chemical structures for the compounds with the pyran group as a spacer

The interest of the comparison between series **17** and series **18** lies in the substitution of the acceptor group, that is, in **17** the tricyan acceptor has a phenyl group, however, in **18**, it has an amine group, which is a stronger electron donor group. The substitution of the amine for the phenyl group leads to a clear hypsochromic shift of the absorption λ_{max}, and a clear decrease of the NLO response.

Finally, with all the studied compounds we can compare the efficiency of the pyran group as a donor with that of dithiol. The compounds **13** and **7** share the phenyloxazolone acceptor with **16** and the tricyanopyran with **19**, respectively. As can be seen, the dithiol group yields much larger $\mu\beta$ values, showing the limitations of pyran acting as a donor.

3.2.3.2 As a Spacer Group

In the following series of compounds, the pyranylidene, to which an acceptor group is linked, behaves strictly as a spacer. In order to study the NLO response–transparency relationship, different 1D (donor–acceptor) and V-shaped 2D (donor–acceptor–donor) geometries with multiple donors and acceptors placed in different ways are proposed [19]. On the other hand, as has been mentioned before, the linear polyenic spacers show a limited thermal and photochemical stability. The 1D and 2D structures combined with several acceptor groups are depicted in Fig. 3.10.

At this point, it is good to remember that by using the EFISH experiment there is only access to a part of the NLO response, as we measure the scalar product of the molecule's dipole moment in the ground state and the vector part of the first-order hyperpolarizability tensor β. In 2D molecules there can be important components of the second-order tensor that are not parallel to the dipole moment μ so by using EFISH, the characterization of the molecular nonlinearity is not complete.

The absorption results and the $\mu\beta$ values are shown in Table 3.8. All the compounds show positive solvatochromism. The order followed by the absorption

Table 3.8 Absorption-maximum wavelengths and hyperpolarizabilities, measured in CH_2Cl_2, of the compounds with a pyran group as a spacer

	λ_{max} (CH_2Cl_2)	λ_{max} (DMSO)	$\mu\beta[10^{-48}esu]$	$\mu\beta$ (0) $[10^{-48}esu]$
20a	492	498,557(s)	1440	990
20b	526	535	1570	1010
20c	530,598(s)	541,618(s)	1500	945
20d	533	540	2300	1455
20e	589	598	3000	1700
20f	649	654	2200	1045
21a	478	498	890	625
21b	520	533	900	585

maximum wavelengths is **20f** > **20c** > **20e** > **20d** > **20b** > **20a**. The structure with TCF acceptor group is the one showing absorption closer to the red, but it has to be remembered that the compound has an additional double bond with respect to the rest. The most efficient derivative is **20e**, which contains tricyanophenyl as an acceptor group, followed by **20d** and **20f**, which show hardly any differences. Experimentally, it is not possible to tell **20a** from **20b** and **20c**. In fact, in the analogous linear compounds **21a** and **21b** there is no difference between the contribution of the dicyano and indanedione groups.

On the other hand, it seems that the 2D V-shaped compounds show a shift to the red with respect to their homologous 1D structures, giving rise, at the same time, to better NLO results. The achieved results appear to show that the pyranylidene group works as a polyenic spacer and not as an auxiliary donor group.

3.2.3.3 As a Donor and Spacer Group

Continuing with the research on the incorporation of pyran derivates in chromophores for NLO, a group of molecules where this group appears in two different positions is presented in Fig. 3.11. In one of them it acts as a spacer and in the other one as a donor. Also in this group, 1D and 2D structures have been studied [20].

In the previous section, we have seen, by using two examples, that the 2D compounds with an extra donor group give rise to a larger $\mu\beta$ than their linear equivalent with just one donor group. In this particular case (Fig. 3.11), as we will see below, the differentiation is not so clear. The absorption results show a clear bathochromic shift of the maximum on passing from two double bonds in the linear polyenic chain to three (from **22** to **24**), as has happened so far when lengthening the chain. Chain lengthening by incorporating the pyran group in the spacer also results in a shift to the red, though in this case it is much smaller. On the other hand, the increase of the acceptor strength (**c** > **b** > **a**) causes a shift of the absorption maximum to lower energies. All the compounds show positive solvatochromism when dichloromethane is replaced by a more-polar solvent such as DMSO.

Fig. 3.11 Chemical structures of the compounds with pyran as donor and spacer group

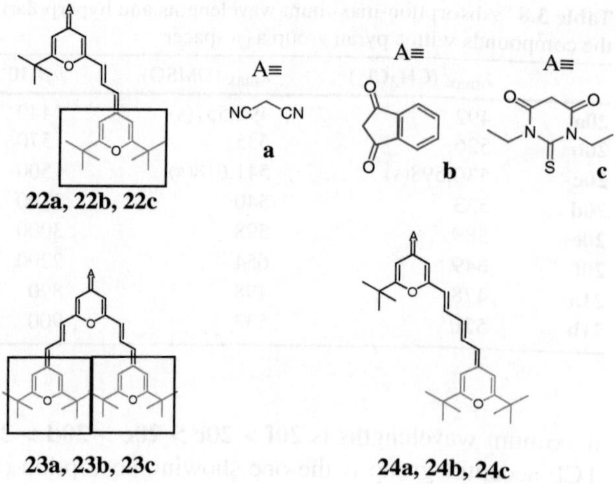

With regard to NLO properties, the expected trend cannot be seen when the strength of the acceptor group is enhanced in any of the three compounds, since the **b** ones, with indanedione, show the lowest $\mu\beta$ value. The thiobarbituric group acts as a better acceptor than the dicyano, giving rise to larger $\mu\beta$ values. Since the two-level model approximation is valid only for dipole rod-like compounds, with only a low-energy relevant transition, the $\mu\beta(0)$ values for compounds **22** will be only approximate. The geometry of compounds **23** precludes giving an estimate of the corresponding $\mu\beta(0)$ value. The addition of a double bond in the linear conjugated chain doubles the NLO response. On passing to the 2D structure a clear improvement of the NLO response cannot be seen. For the thiobarbituric acceptor the change in the geometry increases the $\mu\beta$ value by approximately 30 %, but in the other two cases the results are worse.

In the previous section (pyran as spacer) when comparing the 1D compounds with the 2D ones, we found the opposite phenomenon. Thus, when going from compounds **21** (1D) to **20** (2D) there was an increase of the NLO response. On the other hand, if we compare the linear compounds with different donors, diethylaniline (**20**) and pyranylidene (**22**), the donor character seems to be very similar. However, when comparing both V-shaped chromophores, **20** and **23**, the $\mu\beta$ values are much larger for diethylaniline.

The difficulty in understanding the results obtained comes probably from the two-dimensional character of the nonlinearity of these compounds. The calculations made by the group of Prof. Garín using the coupled perturbed Hartree–Fock (CPHF) method and time-dependent density functional theory (TDDFT) [16, 17] show the importance of the magnitude of the "diagonal" β_{zzz} and "off-diagonal" β_{zyy} components. It makes clear the usefulness of making different kinds of measurements to have access to other components of the hyperpolarizability tensor.

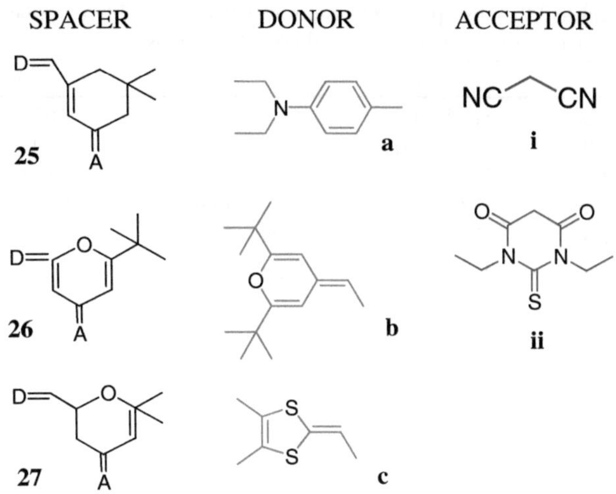

Fig. 3.12 Structures of the compounds with isophorone, pyran and dihydorpyran spacers

3.2.4 *Merocyanines with Isophorone, Pyran and Dihydropyran Spacers*

In the next group of molecules, the study will be comparative between different cycled spacers: isophorone, pyran and dihydropyran. At the same time, the effect of three different donor groups will be analysed: diethylaniline, *tert*-butylpyran and dithiol, and two acceptors: thiobarbituric and dicyan [10]. The compounds measured have been gathered according to the spacer group, in the following way:

With isophorone spacer: **25ai, 25aii, 25bi, 25bii, 25ci, 25cii**
With pyran spacer: **26ai, 26aii, 26bi, 26bii, 26ci, 26cii**
With dihydropyran spacer: **27ci, 27aii**

a, **b** and **c** are different donor groups and **i** and **ii** the acceptor ones, as shown in Fig. 3.12.

As Fig. 3.12 shows, the compound **26ai** is the same as **21a**, which was studied in the previous section.

The results of the absorption-maxima wavelengths in two different solvents, as well as the $\mu\beta$ values obtained by EFISH, are shown in Table 3.9.

All the compounds show positive solvatochromism. The chromophores with isophorone (**25**) show a shift to lower energies in the absorption maximum, with respect to the ones containing pyran-derived spacers (**26**, **27**), in both solvents. Of these two, the spectrum of the chromophore with dihydropyran (**27**) shows a bathochromic shift with respect to the spectrum of compounds **26**.

Regarding the NLO response, the merocyanines with isophorone as a spacer show larger $\mu\beta$ values than the ones containing pyran. Considering that the latter

Table 3.9 Absorption-maximum wavelengths and hyperpolarizabilities, in CH_2Cl_2, of the compounds with pyran as donor and acceptor groups

	λ_{max} (CH_2Cl_2)	λ_{max} (DMSO)	$\mu\beta[10^{-48}esu]$	$\mu\beta$ (0) $[10^{-48}esu]$
22a	519,558(s)	526	1070	700
22b	547,557(s)	559	860	530
22c	587	593	1160	650
23a	522,574(s)	527,582(s)	900	–
23b	552,614(s)	560,631(s)	670	–
23c	552,633	561,658	1500	–
24a	543	555	2630	1633
24b	572	587	1600	932
24c	602	619	2530	1370

Table 3.10 Absorption-maximum wavelengths and hyperpolarizabilities, in CH_2Cl_2, for the compounds with isophorone, pyran and dihydropyran spacers

	λ_{max} (CH_2Cl_2)	λ_{max} (DMSO)	$\mu\beta[10^{-48}esu]$	$\mu\beta$ (0) $[10^{-48}esu]$
25ai	519	538	1150	750
25bi	561	579	1430	850
25ci	560	569	1590	950
25aii	594	620	2500	1410
25cii	638	661	2315	1135
26ai	478	498	890	625
26bi	519,558(s)	526	1070	698
26ci	518	530	570	370
26aii	537	553	1220	765
26cii	575	585	1080	625
27ai	516	529	1040	682
27bi	565,605(s)	616	1000	590
27aii	599	570,614(s)	1850	1010

can withstand higher temperatures, it would be interesting to explore the resultant hyperpolarizability of the same compounds with the dihydropyran spacer. However, as can be seen in Table 3.10, changing the pyran spacer did not give rise to a related increase of the nonlinear response: the trend depended on the acceptor.

Changing the dicyan acceptor group (**i**) to a thiobarbituric one (**ii**) is advantageous in all cases, giving larger $\mu\beta$ values for all the pairs: **25aii** > **25ai**, **26aii** > **26ai**, **25cii** > **25ci**, **26cii** > **26ci** and **27aii** > **27ai**. As for the donor groups, despite the fact that it seems that the diethylaniline group is a better donor than the dithiol one, this does not happen for the compound with isophorone and dicyano (**25ai**), so there is no clear trend. On the other hand, the pyran donor shows similar or better results than the dicyan.

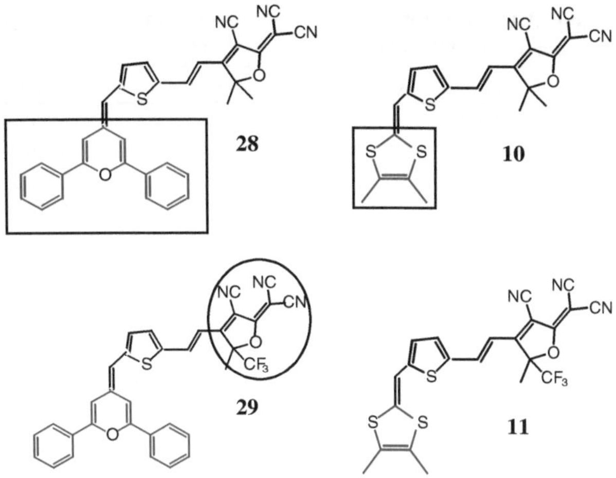

Fig. 3.13 Chemical structures of the compounds with thiophene spacers, different donor and acceptor groups and same chain length

3.2.5 Merocyanines with Thiophene Spacers: Influence of Chain Lengthening on Both Sides of the Thiophene

In the next group of molecules, NLO merocyanines with polyenic spacers containing thiophene will be analysed. In addition to a study about different donors and acceptors, the analysis of the influence of the lengthening topology of the spacer (polyenic chain) has been included. The structures of the molecules are shown in Fig. 3.13. As can be seen, some of the structures (**10**, **11**) were part of the study presented in Sect. 3.2.2.

First of all, a comparative study of two acceptor groups will be made, TCF and TCF-CF_3, since they have given rise to the compounds with the largest $\mu\beta$ values. In Sect. 3.2.2.1, it was shown that this substitution hardly influenced the NLO properties. However, there are publications in which it is assured that the trifluoromethane significantly improves the NLO properties [21]. Regarding the donor groups, we have diphenylpyran and dithiol for now.

In Table 3.11 the values obtained for the absorption maximum and the molecular hyperpolarizabilities of the mentioned molecules are summarized. The compounds with TCF, **28** and **10**, show negative or zero solvatochromism, which means that they would be close to the CL. However, with the trifluoromethylene group, **29** and **11**, the solvatochromism is clearly positive, and $\mu\beta$ doubles its value. Besides, the values for the diphenylpyran, **28** and **29**, are larger than those achieved by the compounds with dithiol, **10** and **11**.

The analysis is extended to compounds that show the same structure as the previous ones except for the spacer, which includes one extra double bond, as can be seen in Fig. 3.14. In this figure, it is shown that the length of the spacer is the same in all the chromophores, but the lengthening of the polyenic chain takes place on the acceptor side in **30**, **31** and **32**, and on the donor side in **12**, **33** and **34**.

Table 3.11 Absorption-maximum wavelengths and hyperpolarizabilities, in CH_2Cl_2, of the compounds with thiophene spacer, different donors and acceptors, and the same chain length

	λ_{max} (CH_2Cl_2)	λ_{max} (DMSO)	$\mu\beta[10^{-48}$esu]	$\mu\beta$ (0) [10^{-48}esu]
28	708	708	6790	2630
10	723	711	5440	1980
29	827	847	13600	2735
11	793	815	11200	2850

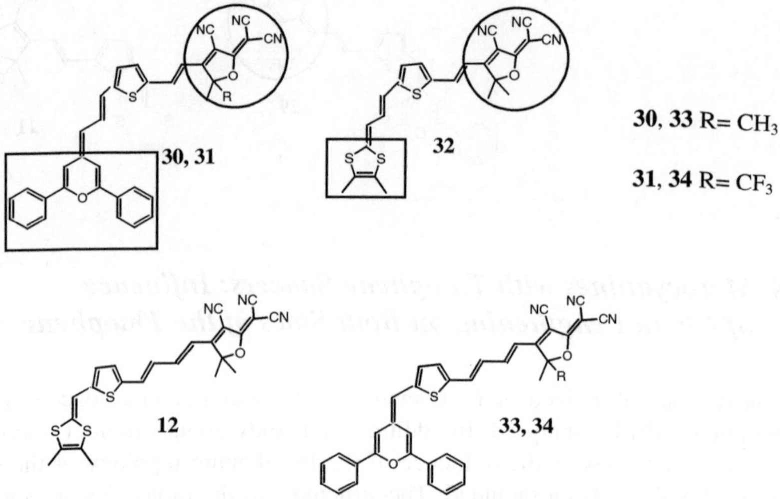

Fig. 3.14 Chemical structures of the compounds with thiophene spacer and an extended polyenic chain on one side or the other

In Table 3.12 the influence of these design differences on the NLO properties and on the absorption-maximum wavelengths is shown. According to these results, the solvatochromism for these compounds with the diphenyl pyran group is clearly positive, while for **12** and **33**, with dithiol, it seems to be negative, which would place it closer to the CL (region C in Marder's curve) and gives rise, again, to lower $\mu\beta$ values than in the case of the diphenylpyran. Moreover, it can be checked, just as in the short chain compounds, that the most efficient acceptor would be TCF-CF_3, as $\mu\beta$ for **31** and **34** reaches values more than double that for **30** and **33**.

In the dithiol compounds (**32** and **12**) as well as in the ones bearing diphenylpyran as a donor (**30**, **31**, **33** and **34**), the lengthening of the spacer on the acceptor side, **30**, **31**, doubles the $\mu\beta$ value as compared with the same compound lengthened on the donor side, **33** and **34**.

Globally speaking, the largest result so far is that for compound **34**, lengthened on the acceptor (TCF-CF_3) side and with diphenylpyran as a donor group. It is interesting to note that the $\mu\beta$ values for **30**, **31**, **33** and **34** are influenced by

Table 3.12 Absorption-maximum wavelengths and hyperpolarizabilities, in CH_2Cl_2, of the compounds with a thiophene spacer and the polyenic chain lengthened on one side or the other

	λ_{max} (CH_2Cl_2)	λ_{max} (DMSO)	$\mu\beta[10^{-48}esu]$	$\mu\beta$ (0) $[10^{-48}esu]$
30	739	742	11940[a]	4050
31	845	857–951	35000[a]	6040
32	719	691	5600	2100
12	716	–	12080	4470
33	734	703	17300[a]	6000
34	827–903	842–944	37000[a]	7500

[a]corrected values assuming the absorption at 954 nm

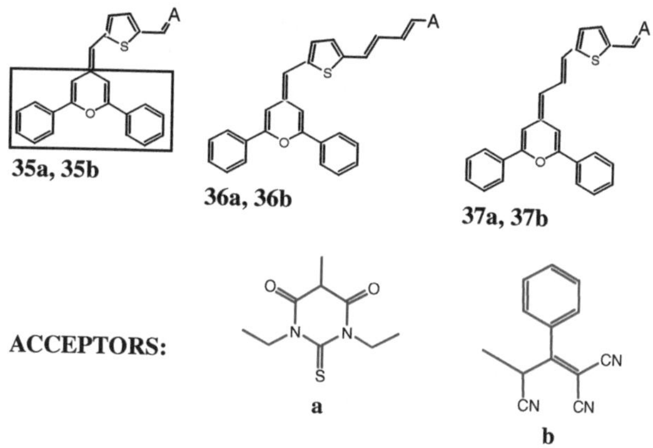

Fig. 3.15 Chemical structures with thiophene spacer and lengthening of the polyenic chain on one side or the other, containing diphenylpyran as a donor group

the absorption that these molecules show at 954 nm. Very dilute solutions have been measured and the $\alpha_{2\omega}$ coefficients have been considered. Thus, compounds **31** and **34** needed the largest correction in their measured values. On the other hand, the absorption proximity of the compound to the second harmonic wavelength (954 nm) induces a very large NLO response, as it is increased by resonance. Usually, this is corrected by using the two-level model and the wavelength of the absorption maximum, which enables us to calculate the static $\beta(0)$ coefficient. However, this correction does not allow reliable values to be obtained in cases such as the one we are dealing with, in which the compound shows a very high absorption ($\varepsilon^\circ \approx {}^\circ 6.6\ l\ mol^{-1}\ \mu m^{-1}$) at the second harmonic wavelength.

In order to check that lengthening on the acceptor side gives better results, other structures with diphenylpyran as a donor group and different, less efficient, acceptor groups, such as thiobarbituric and tricyanophenyl, were measured. These compounds are detailed in Fig. 3.15.

In Table 3.13 the NLO properties of these molecules measured in dichloromethane are shown.

The solvatochromism is positive for all the compounds, though quite small. The acceptor change from thiobarbituric to tricyanophenyl (from **a** to **b**) shifts the band to the red in the three studied cases, and the $\mu\beta$ values are higher for

Table 3.13 Absorption-maximum wavelengths and hyperpolarizabilities, in CH_2Cl_2, of the compounds with thiophene spacer, diphenylpyran donor and different chain lengthening

	λ_{max} (CH_2Cl_2)	λ_{max} (DMSO)	$\mu\beta[10^{-48}$esu]	$\mu\beta$ (0) $[10^{-48}$esu]
35a	644	646	1380	670
35b	676	677	2900	1260
36a	688	696	3350	1400
36b	731	726	8500	2990
37a	668	679	3130	1400
37b	715	717	6900	2600

tricyanophenyl. In both cases, the chain lengthening causes a bathochromic shift and a noticeable improvement in the nonlinear response (**36**,**37** > **35**). Again, it can be seen that a lengthening on the acceptor side doubles the $\mu\beta$ value, according to the previously observed displacements in the absorption and confirming the trend already described for the compounds with dithiol and diphenylpyran as donors.

3.3 Conclusions

Below, the most relevant conclusions of the study of the NLO molecular response developed in this chapter are summarized.

The results for the $\mu\beta$ values obtained by EFISH at two different wavelengths gave rise to a good agreement of the $\mu\beta(0)$ values in compounds bearing the azobenzene group. This result proves the validity of the two-level model for this kind of dipole rod-like compounds. The results obtained for different substituents show the benefits of increasing the strength of the donor and acceptor groups in structures of this kind. On the other hand, the formation of 2D structures, through the lengthening of these structures by using aromatic polyenes in the acceptor ending, leads to slightly higher $\mu\beta$ values.

Regarding the series of merocyanines studied, the lengthening of the spacer chain has given positive results throughout all the research in terms of increasing the efficiency of the nonlinear response, though it also causes bathochromic shifts. In addition, in general, it has been confirmed that the proaromatic character of both the donor and the acceptor groups introduces an outstanding improvement in their properties.

The incorporation of the thiazol group resulted in a series of merocyanines that are located at the RHS of Marder's curve. The enlargement of the chain in these compounds places them progressively closer to region D, reaching very large negative $\mu\beta$ values. On the other hand, substitution of different donor groups allows the NLO response to be modulated, giving rise to systems that would be assigned to region B/C of the curve. As for the study of the influence of the morphology of the spacer, it has been seen that the introduction of cycles to the chain to improve the stability with respect to thermal or photo-oxidative degradation does not imply, in general, an increase of the response, different trends having been observed when using a thiophene group ($\mu\beta$ decreases) or the isophorone one, which modifies $\mu\beta$ in both ways. However, the inclusion of pyranylidene derivatives as donor and spacer simultaneously provides 2D structures showing a noticeable improvement in the NLO response. If the pyran-derived spacer is compared with the isophorone one, the molecules containing the latter offer a larger NLO response, but their thermal stability is clearly lower. Using the dihydropyran spacer, which would be an intermediate moiety between the last two, did not improve the NLO response relative to the pyran. Finally, it has also been shown that lengthening the polyene spacer chain on the side of the acceptor group remarkably improved the NLO response in the series of compounds with thiophene in the conjugated chain.

Concerning the acceptor groups, the best performance is offered by TCF substituted with trifluoromethylene, although in the merocyanine with dithiol as donor there was no improvement in this regard. However, it was found that the compounds were in region B/C, since a change to a more polar solvent resulted in negative values of the hyperpolarizability. In the study of merocyanines with thiophene spacer, the largest positive NLO response was found for the compounds in which the spacer was lengthened on the acceptor side. The effect that causes the change of the substituent in the acceptor group was also studied, replacing an aromatic group by an electron-donor one, which gave rise to a lower value of $\mu\beta$.

As for the donor groups, a study of the effect of a variety of them in different molecular structures was carried out. It is concluded that the best results in terms of $\mu\beta$ values were obtained with diphenylpyran.

Finally, given the high stability and good response of the compounds with thiophene in the conjugated chain lengthened on the acceptor group side, they were selected for incorporation into high-T_g polymers, and the studies in "guest–host" systems began. In addition, side-chain polycarbonates have been prepared with some of these compounds, particularly with **28**, containing diphenylpyran and TCF.

References

1. L.R. Dalton, P.A. Sullivan, D.H. Bale, Electric field poled organic electro-optic materials: state of the art and future prospects. Chem. Rev. **110**(1), 25 (2010)
2. M.J. Cho et al., Recent progress in second-order nonlinear optical polymers and dendrimers. Prog. Polym. Sci. **33**(11), 1013 (2008)
3. Y.-J. Cheng et al., Large electro-optic activity and enhanced thermal stability from diarylaminophenyl-containing high-β nonlinear optical chromophores. Chem. Mater. **19**(5), 1154 (2007)
4. R. Alicante et al., Synthesis and nonlinear optical properties of side chain liquid crystalline polymers containing azobenzene push-pull chromophores. J. Polym. Sci., Part A: Polym. Chem. **48**(1), 232 (2010)
5. N. Tirelli et al., Structure-activity relationship of new organic NLO materials based on push–pull azodyes. 1. Synthesis and molecular properties of the dyes. J. für. praktische. Chemie. **340**(2), 122 (1998)
6. I.C. Pintre et al., Bent-core liquid crystals in a route to efficient organic nonlinear optical materials. J. Mater. Chem. **20**(15), 2965 (2010)

7. M.S. Wong et al., Non-classical donor-acceptor chromophores for second order nonlinear optics. Adv. Mater. **8**(8), 677 (1996)
8. M. Yang, B. Champagne, Large off-diagonal contribution to the second-order optical nonlinearities of Λ-shaped molecules. J. Phys. Chem. A. **107**(19), 3942 (2003)
9. I.C. Pintre et al., Liquid crystalline and nonlinear optical properties of bent-shaped compounds derived from 3,4′-biphenylene. J. Mater. Chem. **17**(21), 2219 (2007)
10. K.D. Singer et al., Second-order nonlinear-optical properties of donor- and acceptor-substituted aromatic compounds. J. Opt. Soc. Am. B. **6**(7), 1339 (1989)
11. R. Andreu et al., Isophorone- and pyran-containing NLO-chromophores: a comparative study. Tetrahedron Lett. **51**(28), 3662 (2010)
12. R. Andreu et al., Aromatic/proaromatic donors in 2-dicyanomethylenethiazole merocyanines: from neutral to strongly zwitterionic nonlinear optical chromophores. Chem.: Eur. J. **17**(3), 826 (2011)
13. R. Andreu et al., Decreased optical nonlinearities upon CF3 substitution on tricyanofuran acceptors. Org. Lett. **10**(21), 4963 (2008)
14. A.J. Kay et al., Synthesis and linear/nonlinear optical properties of a new class of 'RHS' NLO chromophore. J. Mater. Chem. **14**(8), 1321 (2004)
15. M. Barzoukas et al., Very large quadratic nonlinearities in solution of two push-pull polyene series: effect of the conjugation length and of the end groups. Chem. Phys. **133**(2), 323 (1989)
16. S. Alías et al., Synthesis, characterization and optical properties of merocyanines derived from malononitrile dimer. Tetrahedron Lett. **48**(37), 6539 (2007)
17. S. Alías et al., Iminium salts of ω-dithiafulvenylpolyenals: an easy entry to the corresponding aldehydes and doubly proaromatic nonlinear optic-phores. J. Org. Chem. **73**(15), 5890 (2008)
18. R. Andreu et al., 4*H*-pyran-4-ylidenes: strong proaromatic donors for organic nonlinear optical chromophores. J. Org. Chem. **74**(17), 6647 (2009)
19. R. Andreu et al., New one- and two-dimensional 4*H*-pyranylidene NLO-phores. Tetrahedron Lett. **50**(24), 2920 (2009)
20. R. Andreu et al., Linear and V-shaped nonlinear optical chromophores with multiple 4*H*-pyran-4-ylidene moieties. J. Org. Chem. **75**(5), 1684 (2010)
21. S. Liu et al., Focused Microwave-Assisted Synthesis of 2,5-Dihydrofuran Derivatives as Electron Acceptors for Highly Efficient Nonlinear Optical Chromophores. Adv. Mater. **15**(7–8), 603 (2003)

Chapter 4
Piperazine Azopolymer Thin Films

This chapter describes the linear and nonlinear optical (NLO) properties of thin films prepared from various liquid crystalline (LC) azo polymers with different donor groups and different azo chromophore content. Initially, freshly made samples are characterized by refractive index measurements and optical absorption. These films are then subjected to various treatments, both thermal and irradiation with light, during which the evolution of these properties is followed so as to investigate the order acquired by the chromophores. Non-centrosymmetric order is induced by applying a corona electric discharge and the nonlinear response is studied by second harmonic generation (SHG) measurements. The influence of previous treatments with light and high temperature on the mentioned response is also analysed. Likewise, some methods for optimizing the second harmonic signal of these materials are proposed, and both the temporal and thermal stability of the second harmonic signal are studied.

We begin by discussing the results obtained for the homo polymers, and in a second part the behaviour of the copolymers with different side-chain azo chromophore content will be analysed.

4.1 Homopolymers

The liquid crystalline polymer (LCP) on which this part of the study focuses is a polymethylacrylate whose side-chain is a mesogenic unit consisting of a piperazine donor group attached to the cyano acceptor by an azobenzene group. This side-chain is linked covalently to the polymer through a flexible aliphatic chain of six methylenes. Throughout this work we will refer to this material as Pol-PZ-CN. For the sake of comparison of certain behaviours, a parallel study on a known LCP with identical structure except for the donor group, oxygen in this case, was conducted. This material will be referred to as Pol-O-CN. Two more polymers were synthesized from the azomonomers with nitro and dicyano acceptor groups that were characterized in Chap. 3. However, the low molecular weight achieved in the

R. Alicante, *Photoinduced Modifications of the Nonlinear Optical Response in Liquid Crystalline Azopolymers*, Springer Theses, DOI: 10.1007/978-3-642-31756-9_4,

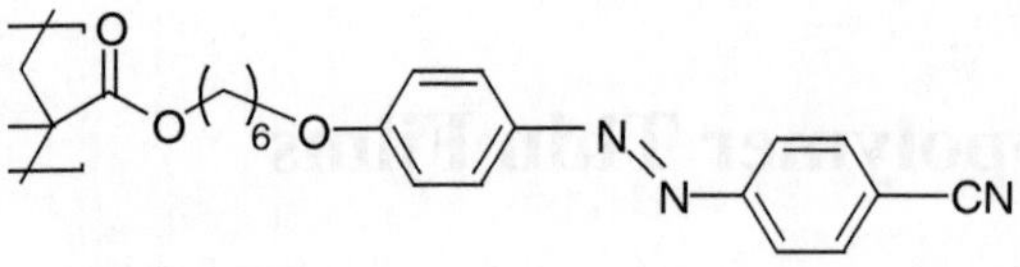

Pol-O-CN

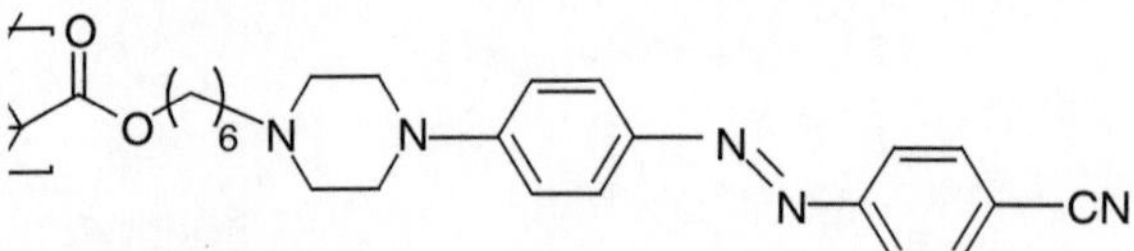

Pol-PZ-CN

Fig. 4.1 Chemical structures of the studied azopolymers

Table 4.1 Thermal properties of the polymers studied

Polymer	TGA decomposition temp. [°C][a]	Phase transitions [°C][b]
Pol-PZ-CN	245	g 77 Sm_A 114/decomp T > 200
Pol-O-CN	280	g 56 Sm_A 163 I/decomp T > 200

[a] Decomposition detected from the thermogravimetric curve
[b] Data corresponding to the first heating scan

synthesis as well as the poor solubility of Pol-PZ-NO_2 and Pol-PZ-DCN precluded the preparation of films of sufficient quality to study their optical properties.

The chemical structures of the polymers studied are shown in Fig. 4.1. Both LCPs incorporate a relatively strong electron acceptor group, CN, and a different donor group in each case, as we have said.

The piperazine group in Pol-PZ-CN has a higher electron-donating character than the O in Pol-O-CN, meaning that we can expect a larger macroscopic nonlinear response of the former.

The thermal characterization of the two polymers is shown in Table 4.1. The glass transition temperature T_g, as shown in the table, is greater for Pol-PZ-CN, resulting in an increased mechanical and thermal resistance of this material compared to Pol-O-CN. In addition, both exhibit thermotropic LC behaviour with smectic phases. Moreover, the Pol-PZ-CN polymer decomposes at high temperatures before reaching the isotropic state, while the Pol-O-CN can indeed reach that state.

4.1.1 *Optical Study*

4.1.1.1 Characterization of As-Prepared Films

Pol-PZ-CN polymeric films about 0.6 m thick were prepared by casting a solution of the polymer in dichloromethane onto fused silica substrates. The absorption

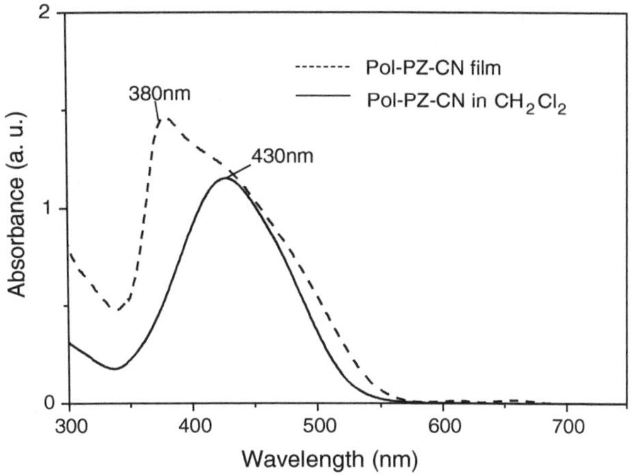

Fig. 4.2 Absorption spectra at normal incidence with non-polarized light, of the Pol-PZ-CN polymer in dichloromethane solution and as a freshly made film

spectra of a solution of Pol-PZ-CN and of a freshly made film are shown in Fig. 4.2.

In both spectra there is a broad band, but they show markedly different shape and position. The band obtained in solution is more symmetrical, with the maximum around 430 nm. As explained in Chap. 1, it corresponds to the characteristic absorption of *pseudo-stilbene*-type compounds, in which the bands of n-π* and π-π* transitions of the *trans* isomers appear overlapped. Since the solution is very dilute, we can expect that this band corresponds mainly to isolated azo chromophores. On the other hand, in the case of the films the overlap of several bands is evident, with the most intense peak around 380 nm. The shape shown in Fig. 4.2 is characteristic of systems in which there is aggregation. In particular, several authors have related similar absorption spectra (with the maximum in the high-energy region) to a significant presence of H-type aggregates, described in Chap. 1 [1]. Although it is not correct to speak of global displacements of the absorption, because they are complex bands, the presence of H-type aggregates causes the displacement of the absorption to higher energies [2–6]. On the other hand, the presence of J-type aggregates, which absorb at higher wavelengths than the isolated chromophores, could explain the widening also observed in the lower energy region, because both spectra comprise complex bands. Therefore, these complex bands are the result of electronic transitions of different species with different oscillator strength: isolated and aggregated azobenzenes.

In Fig. 4.3, a possible deconvolution of the complex spectra is proposed, which in this case would consist of three bands corresponding to the different azobenzene moieties, isolated and aggregated. In the same figure, it can be seen that for a certain ratio of the present species, the sum of the three bands that are proposed reproduces well the experimental data. This has been used by several authors [5, 6] to qualitatively analyse the evolution of the absorption band when the films undergo different treatments that modify the aggregation.

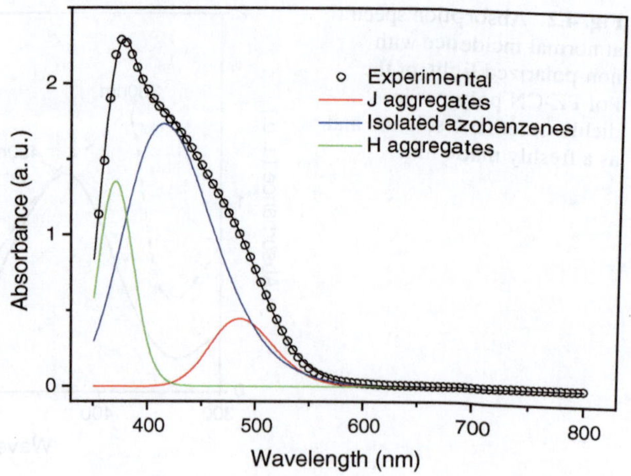

Fig. 4.3 Possible deconvolution of the absorption spectrum for a freshly made film

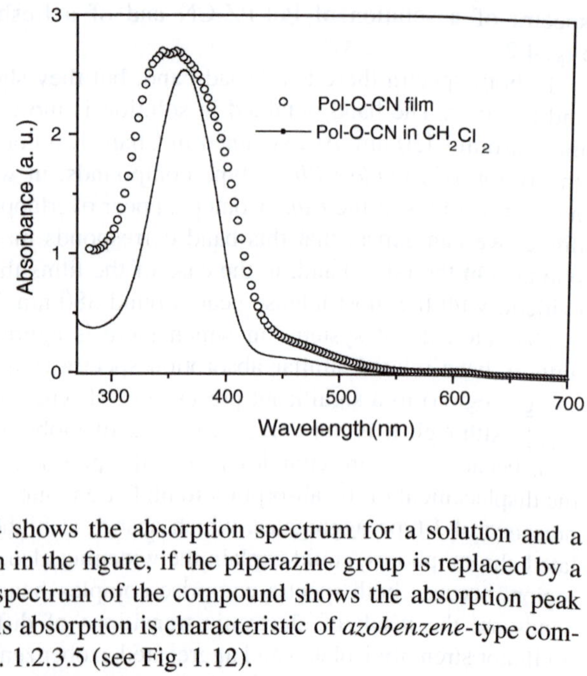

Fig. 4.4 Absorption spectra of Pol-O-CN polymer in dichloromethane solution and in a freshly prepared film, obtained at normal incidence with non-polarized light

For comparison, Fig. 4.4 shows the absorption spectrum for a solution and a film of Pol-O-CN. As shown in the figure, if the piperazine group is replaced by a weaker donor, oxygen, the spectrum of the compound shows the absorption peak shifted towards the blue. This absorption is characteristic of *azobenzene*-type compounds, as described in Sect. 1.2.3.5 (see Fig. 1.12).

On going from the dilute solution to the film, in the case of Pol-O-CN, there is also a broadening and a slight hypsochromic shift of the band, features that can be associated with the existence of different aggregated species, as in the case of Pol-PZ-CN.

In order to obtain some information about how side-chain azo chromophores are arranged in the film, absorption measurements with polarized light and different angles of incidence have been carried out. Next, in Fig. 4.5 the absorption

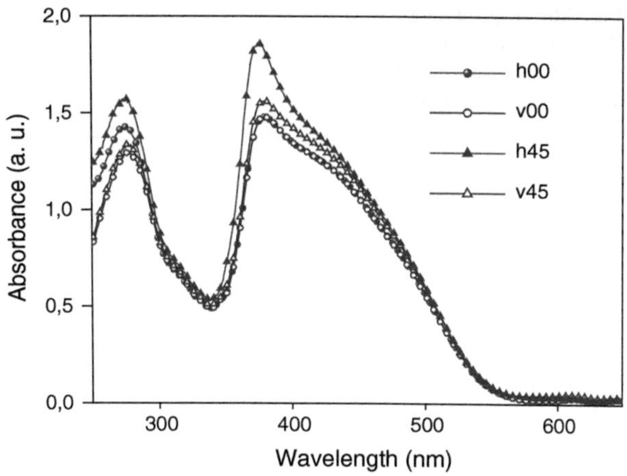

Fig. 4.5 Absorption bands for the Pol-PZ-CN freshly prepared film. (*00* Normal incidence; *45* 45° incidence; *v* vertical polarization of the incident beam; *h* horizontal polarization of the incident beam)

spectrum of a Pol-PZ-CN freshly made film is shown. The spectrum was measured at normal and 45° incidence and with vertical and horizontal polarization.

The absorption enhancement associated with the larger thickness of the sample in the measurements obtained with oblique incidence has been corrected by the Huygens–Fresnel law [4]:

$$\mathrm{OD}_\theta = \mathrm{OD}_0\left[\left(1 - \left(1/n^2\right)\left(\sin^2\theta\right)\right]^{-0.5}$$

where n is the refractive index of the polymer, and OD_0 and OD_θ are the optical densities at normal incidence and at the angle of incidence in question (45° in this case). Even after the correction has been made, the bands at oblique incidence are still more intense than at normal incidence, suggesting some out-of-plane anisotropy in the starting film. The chromophores show homeotropic order preferably, i.e., perpendicular to the plane of the film.

The bands measured for the two polarization directions v00 and h00, at normal incidence, are the same, indicating an isotropic arrangement in the plane of the film. However, an increase of the peak at 380 nm is observed for horizontal polarization (h00) with respect to vertical polarization (v00), at oblique incidence. Since we associate this peak with aggregation, this increase will be due to a greater number of aggregates that are placed perpendicular to the film surface. It seems, therefore, that these aggregates tend to orient homeotropically in the film.

In order to complete the optical characterization of the films, the refractive indices were measured, which provides a direct measurement of the optical anisotropy of the films. As explained in Chap. 2, in the measurement configuration of the films for the transverse electric transmission mode, TE, the guided incident light is polarized parallel to the film plane, and for the magnetic one, TM, perpendicular to the film. The way the coordinate axes in the film are defined (see Fig. 4.6) implies that the optical axis has to be the Z one, so the ordinary index n_o (see Chap. 1) corresponds to light polarization parallel to the XY plane, and the extraordinary one to the polarization parallel to the Z-axis.

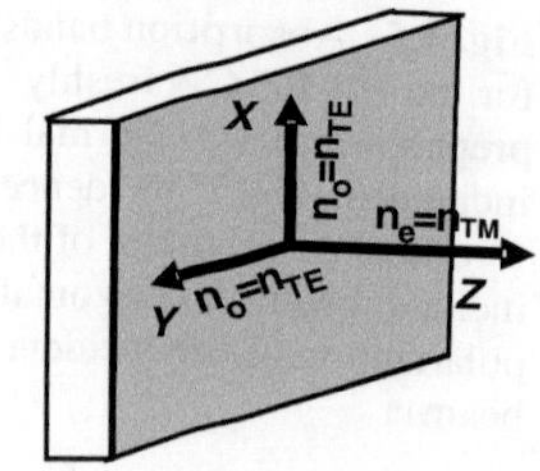

Fig. 4.6 (Figure 2.6) Definition of the film axes in the polarization directions of the transmission modes

Table 4.2 Refractive indices at 633 nm for the as-prepared films

Sample	n_o	n_e	$\Delta n = n_e - n_o$	n_{av}
Pol-PZ-CN	1.63	1.72	0.09	1.66
Pol-O-CN	1.64	1.65	0.01	1.64

The refractive index values measured at 633 nm for the two freshly prepared samples are gathered in Table 4.2. These measurements confirm that the Pol-PZ-CN freshly prepared samples are isotropic in the plane of the film, with $n_x = n_y \equiv n_o$, while the component $n_z \equiv n_e$ is clearly higher. This result agrees with that observed in the optical absorption measurements, indicating a preferred out-of-plane orientation, perpendicular to the film. The starting birefringence $\Delta n = n_e - n_o$, varies slightly from film to film, with their average value at 633 nm near $\Delta n = 0.1$. In the case of Pol-O-CN polymer, a slight homeotropic tendency is also detected, although the measured values of the birefringence hardly exceed $\Delta n = 0.01$.

In addition to providing information on the molecular order in the films, the refractive index measurements are needed to calculate the indices corresponding to the excitation frequency (1907 nm) and to the second harmonic one (954 nm), whose values have to be included in the determination of the nonlinear coefficients of the films. From measurements of refractive indices at 633, 780 and 1306 nm and by using the Sellmeier equation

$$n^2(\lambda) = 1 + \frac{A\lambda^2}{\lambda^2 - B},$$

n values have been obtained at the required wavelengths for both polymers (Fig. 4.7).

4.1.1.2 Thermal Treatment: Optical Study of the Thermotropic Effect

In Appendix C of this report it is explained that the anisotropy of the medium in general and LC properties in particular have a significant impact on the degree of polar order that can be established subsequently by thermal poling. As already mentioned, several studies suggest that the use of liquid crystals would lead to an

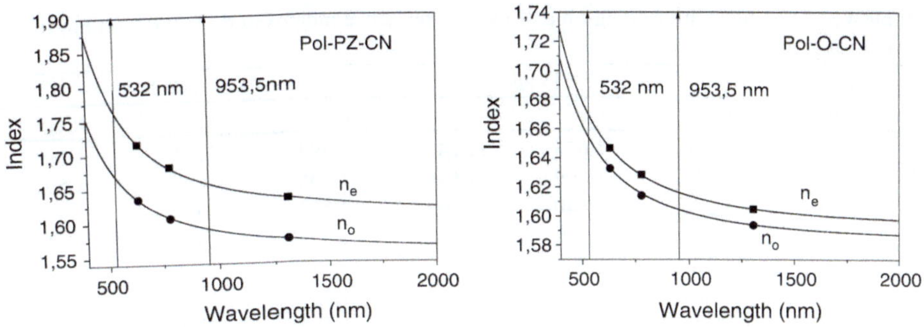

Fig. 4.7 Dispersion of the electric and magnetic refractive indices for Pol-PZ-CN and Pol-O-CN films

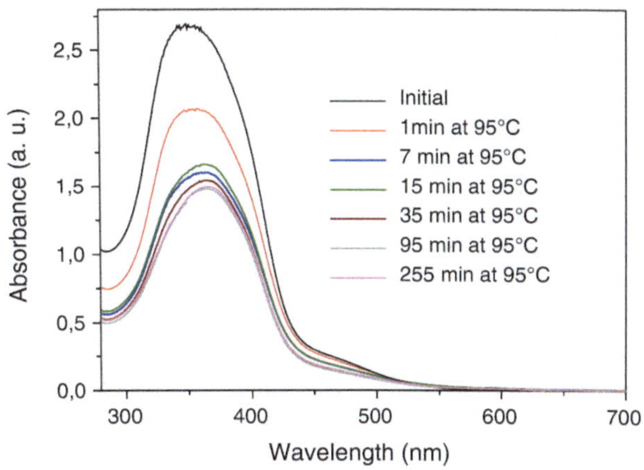

Fig. 4.8 Absorption bands for a Pol-O-CN film under thermal treatment, obtained with non-polarized light at normal incidence

improvement in the stability of the nonlinear response. Moreover, the anisotropic nature would also result in nonlinear coefficient ratios d_{33}/d_{31} higher than 3, which is associated with a better polar order.

Therefore, some studies were conducted on the effect on the anisotropy in freshly prepared films by successively heating them above the mesophase temperature. It is important to recall that most of the processes to break centrosymmetry (electric poling) are performed at high temperature, so it is convenient to know the behaviour of the systems in purely thermal processes before adding other variables.

We will start by showing in Fig. 4.8 the evolution of the absorption spectrum of a ≈ 1 μm thick Pol-O-CN film with time at a temperature of 95 °C. As shown in this graph, when the film is maintained at a high temperature the absorption in the plane of the film decreases, since the material is a thermotropic liquid crystal that is oriented perpendicularly to the surface. The azobenzenes oriented this way do not absorb light at normal incidence.

Table 4.3 Evolution with heating time of the two refractive indices at 633 and 1306 nm for the Pol-O-CN film

Time at 95 °C [min]	633 nm			1306 nm		
	n_o	n_e	n_{av}	n_o	n_e	n_{av}
0	1.636	1.653	1.642	1.594	1.605	1.598
1	1.613	1.709	1.646	1.568	1.640	1.592
3	1.596	1.740	1.645	1.563	1.666	1.598
7	1.587	1.759	1.646	1.559	1.688	1.603
15	1.574	1.788	1.648	1.540	1.689	1.591
35	1.555	1.814	1.646	1.533	1.730	1.601
95	1.549	1.823	1.645	1.522	1.728	1.594
275	1.549	1.826	1.647	1.520	1.727	1.592

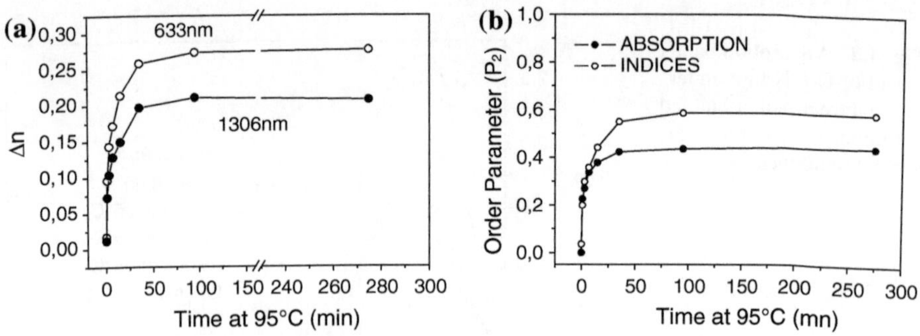

Fig. 4.9 **a** Evolution of the out-of-plane birefringence with heating time in a Pol-O-CN film. **b** Evolution of the order parameter with heating time in a Pol-O-CN film calculated from absorption and refractive index data

The refractive indices of the sample at 633 and 1306 nm were measured at each stage, and the results, together with the average index value, are shown in Table 4.3.

The representation of the evolution of the birefringence with heating time is shown in Fig. 4.9a. As can be seen, the out-of-plane birefringence value increases quickly at first and then more slowly to a saturation value of 0.28, after about 90 min. From these data, we estimated the value of the order parameter P_2, as explained in Appendix C. In Fig. 4.9b the evolution of P_2 with heating time is shown, obtained from both absorption and refractive index measurements. We recall here the expressions used for this purpose.

From the absorption measurements: $P_2 = 1 - \frac{A_\perp}{A_0}$, where A_0 is the initial absorbance of the sample at normal incidence and $A_\perp$ is the absorbance of the axially oriented sample measured at normal incidence.

From the refractive indices: $P_2 = \frac{n_{TM} - n_{TE}}{\delta n_0}$, where n_{TE} and n_{TM} are the electric (polarization parallel to the film) and magnetic (perpendicular) transversal indices, respectively, and $\delta n_0 = \frac{\delta n_{TM} + 2\delta n_{TE}}{3}$. (For this calculation 360 nm is taken as the wavelength value for the absorption maximum.)

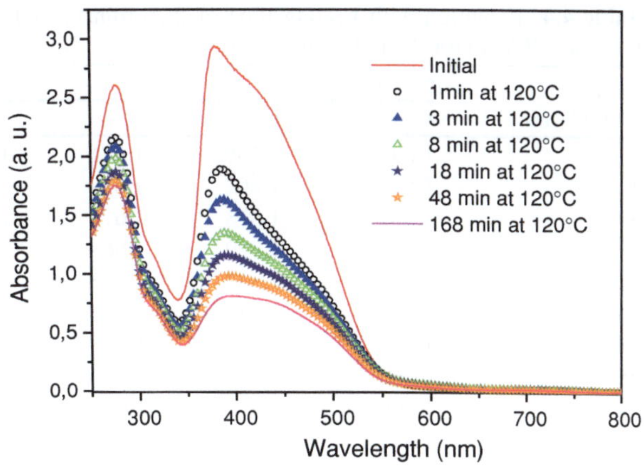

Fig. 4.10 Absorption bands obtained with non-polarized light at normal incidence, for a Pol-PZ-CN film under thermal treatment

From Fig. 4.9b, we can say that the two final results do not converge. This may be due to a deterioration of the films generating an increased light scattering, possibly caused by the accumulated heating time. Therefore, as the heating time increases, the measured $A_\perp$ value would be greater than the real one, and so we would underestimate the value of P_2.

It should be borne in mind that since the initial state is not completely isotropic; the P_2 value obtained by both methods would again be underestimated.

In the case of Pol-PZ-CN the thermal treatment also leads to a decrease of the optical absorption of the films, as shown in Fig. 4.10.

The behaviour of this polymer is similar to that of Pol-O-CN in that the chromophores are oriented perpendicular to the film surface as it is heated. However, in this case, moreover, a change in the shape of the band is evident. This may be related to the various contributions (isolated chromophores and different aggregate types) that we mentioned earlier. This question is discussed in detail below.

In this case, measurements of the evolution of the refractive indices were made at two different temperatures (120 and 150 °C), both above the smectic mesophase. The study at two temperatures will be justified later when we analyse the influence of the temperature used in these polar orientation processes in the presence of an electric field. For the experiment at 120 °C, measurements at 1306 nm were also carried out, which are gathered in Table 4.4.

As an example, Fig. 4.11 shows the light propagation modes (TE electric and TM magnetic) obtained by the guiding light method (Metricon 2010TM) explained in Chap. 2 of this thesis, for a freshly prepared film and a film which has undergone heating at 120 °C for 40 min.

In the figure, the shifts in the minima that occur when the film is heated can be seen. In the case of the TE mode, the minima displace to lower indices and in the TM to higher ones.

Figure 4.12 shows the evolution of the birefringence at 120 (at 633 and 1306 nm) and 150 °C (at 633 nm). As shown, the birefringence at this temperature

Table 4.4 Evolution with heating time of the ordinary and extraordinary refractive indices at 633 and 1306 nm for the Pol-PZ-CN film

Time at	633 nm			1306 nm		
120 °C [min]	$n_x = n_y$	n_z	n_{av}	n_o	n_e	n_{av}
0	1.640	1.730	1.671	1.590	1.661	1.614
1	1.603	1.802	1.672	1.562	1.701	1.610
3	1.595	1.818	1.673	1.556	1.713	1.610
5	1.582	1.829	1.668			
8	1.584	1.837	1.673	1.550	1.728	1.612
15	1.576	1.846	1.671			
18	1.577	1.855	1.675	1.547	1.734	1.612
45	1.566	1.866	1.672			
48	1.566	1.870	1.673	1.539	1.745	1.611
115	1.558	1.879	1.672			
168	1.557	1.892	1.676	1.539	1.770	1.620
954	1.555	1.894	1.676			

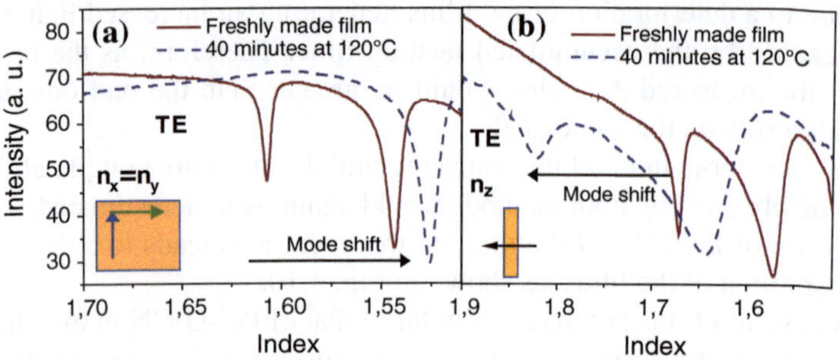

Fig. 4.11 Electric TE (**a**) and magnetic TM (**b**) transmission modes in freshly prepared films and films subjected to heating at 120 °C for 40 min

Fig. 4.12 Evolution of the out-of-plane birefringence with heating time in the Pol-PZ-CN film

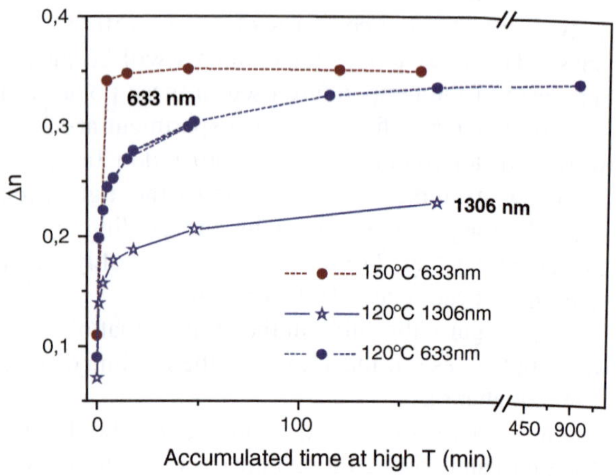

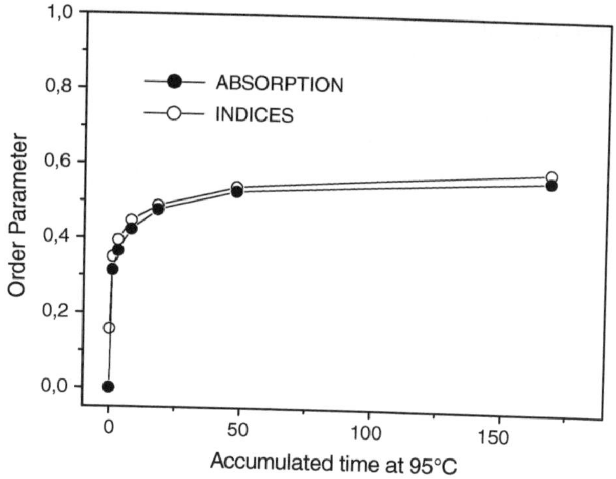

Fig. 4.13 Evolution of the order parameter with heating time for the Pol-PZ-CN film calculated from absorption to refractive index data

follows a similar evolution to that of Pol-O-CN but reaches greater saturation values of about 0.35. After 2 h at 120 °C, the birefringence increases very little. In the case of heating at 150 °C, the initial evolution is much faster, reaching values above 0.3 within 5 min.

As was done for Pol-O-CN, from the results of absorption and refractive index measurements, we calculate the evolution of the order parameter for a film of Pol-PZ-CN, as shown in Fig. 4.13.

In this case, the parameter value calculated from absorption measurements is only slightly lower than that calculated from the evolution of the indices. The good agreement is somewhat surprising, given the complexity of the spectrum that in no way can be assigned to a single moiety. The absorption maximum for this estimate has been placed at 390 nm, which is the part of the band assigned to the H-type aggregated species, so it is assumed that these species play a very important role in the behaviour and alignment of the bulk material. In these Pol-PZ-CN films, the aging caused by thermal treatments is much less visible than in the case of the Pol-O-CN films, and they stay reasonably transparent for much longer.

To conclude the study of the thermally induced anisotropy, we point out that the obtained birefringence values for both materials are stable over time, that is, the homeotropic alignment of chromophores in the side-chain does not decrease significantly over time.

As has been pointed out, aggregation plays a very important role in the optical properties in general, and particularly in the anisotropy of the Pol-PZ-CN polymer. As noted in Fig. 4.10, the freshly prepared film shows signs of a significant degree of aggregation. We have also seen that continued heating leads to changes in the absorption band shape of the spectrum at normal incidence. These changes are reflected more dramatically in the absorption spectrum measured at oblique incidence, as shown in Fig. 4.14 for a sample heated to reach $\Delta n \cong 0.35$.

The figure shows the increase in the overall absorption on varying the angle of incidence, that is, as the sample is rotated around the Z-axis. Moreover, the shape

Fig. 4.14 Absorption spectra, at different incidence angles, of a Pol-PZ-CN film heated to induce maximum out-of-plane birefringence

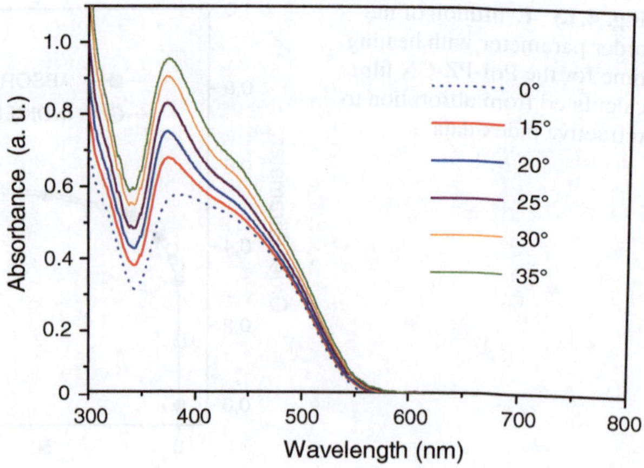

Fig. 4.15 Absorption spectra of a Pol-PZ-CN film at normal and 45° incidence for two mutually perpendicular polarizations of the incident light

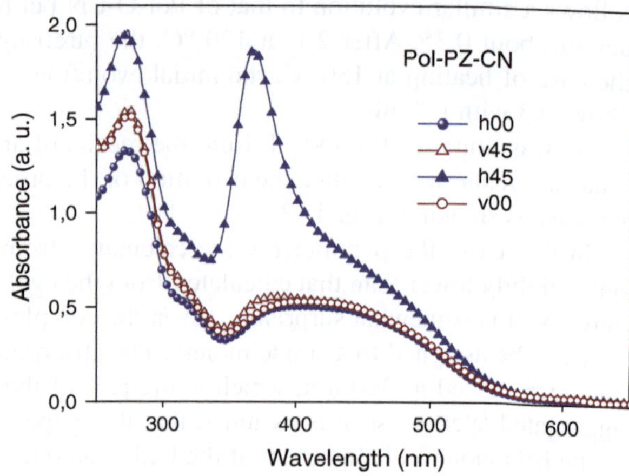

changes significantly, so that the area under the part of the spectrum that is attributed to the aggregates (maximum around 380 nm) becomes more important with increasing angle. This result indicates that the chromophores oriented perpendicularly to the film by the thermotropic effect are largely aggregated.

If we focus our attention now on the absorption spectrum measured at angles of incidence of 0° and 45° of a film under the same conditions as above but using polarized light (horizontally and vertically), this phenomenon is shown most clearly, as can be seen in Fig. 4.15.

The spectra with horizontal (h00) or vertical (v00) polarization are almost identical at normal incidence. However, the spectrum obtained with horizontal polarization and at an oblique direction (h45) shows a significant increase in absorption at 380 nm compared with that obtained with vertical polarization (v45). This is because the aggregation is enhanced at elevated temperatures, placing the

Fig. 4.16 Absorption spectra for Pol-O-CN (a) and Pol-PZ-CN (b) solutions, freshly made, after irradiation, and after a rest period

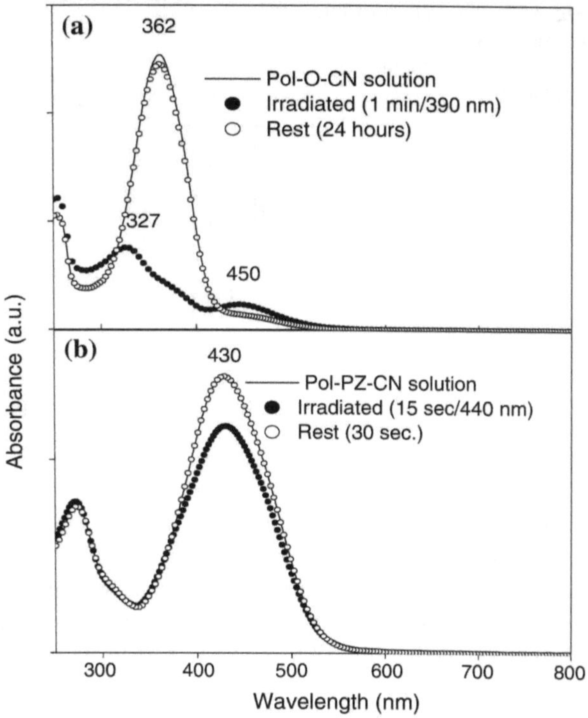

aggregates preferably in the direction perpendicular to the film surface. The relative weight of the area of the band assigned to H aggregates is clearly higher than before the thermal treatment, so the film is in an advanced state of aggregation. Moreover, it can be emphasized that the total area for the heated film is clearly smaller than that for the freshly prepared one, since the absorption oscillator strength corresponding to the aggregated species is much smaller than that for the isolated ones, as indicated in Chap. 1 of this thesis [7]

In the case of Pol-O-CN films, the aggregation effects do not appear so clearly in the absorption spectrum, although their presence is shown in Fig. 4.4.

4.1.1.3 Optical Treatment: UV-Blue Irradiation Effect

Continuing with the effects that the different optical and thermal treatments induce in the polymeric films studied, we will now analyse the influence of illumination with blue/purple light on the optical properties of the films.

This study started with very dilute solutions of both Pol-PZ-CN and Pol-O-CN polymers. They were subjected to brief irradiation with non-polarized light from a 350 W mercury lamp. Bandpass filters of 390 nm/FWHM 100 nm and 440 nm/ FWHM 100 were used. The behaviour of the two polymers under irradiation is very different, as seen in Fig. 4.16.

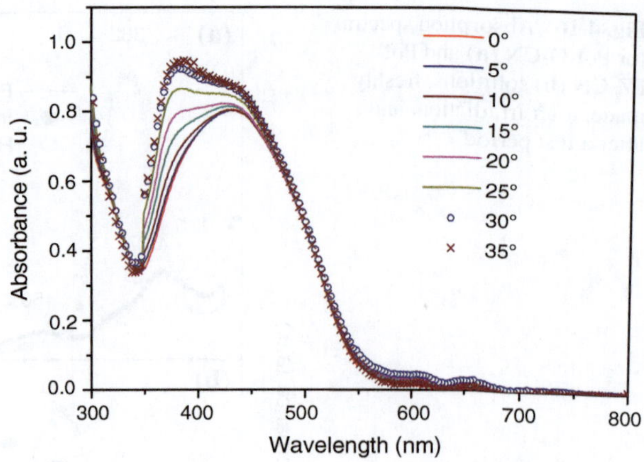

Fig. 4.17 Absorption spectra for the Pol-PZ-CN film first "annealed" and then irradiated

The photoisomerization of the Pol-O-CN sample can be related to an *azobenzene-type* behaviour, thus, when it is irradiated a large decrease in the absorption takes place, due to the *trans* isomers, and the maximum experiences a blue shift, revealing, in turn, a band with a maximum at 450 nm, which is attributed to the *cis* isomers. When the film is left to rest in darkness, the band is restored within 24 h. Moreover, as was pointed out at the beginning of this section, the Pol-PZ-CN spectrum is identified with that of a *pseudoestilbene-type* compound since, upon irradiation, the band of the *trans* azobenzenes decreases but that of the *cis* isomers does not become visible, as the two bands now overlap. This is due to the introduction of a strong donor group, such as piperazine, producing a bathochromic shift of the π-π^* band. As seen in Fig. 4.16b, the recovery of the initial *trans* spectrum, once it is left to rest, is much faster than in the case of Pol-O-CN, so when the strength of the donor substituent on the azobenzene group is increased, the *cis* isomer's half-life decreases [8, 9]. It must be said that the decrease in absorption observed in Fig. 4.16b may actually be larger, since the recovery of the band is very fast and in situ measurements were not made, which would have shortened the time interval between irradiation and the measurement of the absorption spectra. This measurement is just intended to show how fast the characteristic evolution of *pseudoestilbenes* is.

With regard to light irradiation of Pol-O-CN films, the absorption band's behaviour is very similar to that of the solution, and has been described by several authors [10, 11].

However, the case of Pol-PZ-CN polymer is quite different. When a Pol-PZ-CN film is irradiated with light of 440 nm for 1 min at normal incidence, a change in the shape of the absorption band at 0° is obtained, as shown in Fig. 4.17, which is related to the disruption of the aggregates. The use of UV light to break azobenzene aggregates has been widely described in the literature [2, 3] and is based on the fact that the photoisomerization *E-Z* cycles of the azo chromophores destroy the aggregate phase, separating the azo groups.

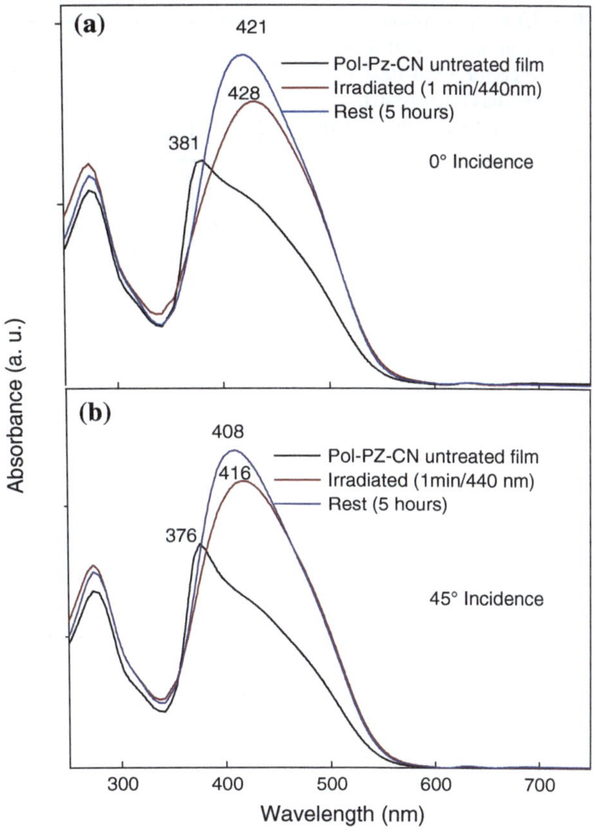

Fig. 4.18 Absorption spectra with non-polarized light for the Pol-PZ-CN film, at 0° (**a**) and 45° (**b**) angles of incidence, irradiated and at rest

In the figure it is seen that the peak at 380 nm has "disappeared" at normal incidence, but it is evident that there is still plenty of aggregated material, which becomes clearer as the angle of incidence of the measurement light increases.

This is why, in order to break the larger number of aggregates, it was decided to irradiate the films at an angle of incidence as large as possible, since illumination in a perpendicular direction would have an effect on only a small proportion, because they are mostly placed homeotropically to the film.

The result of this irradiation can be seen in Fig. 4.18a and b.

It is observed that after the irradiation, the initial band becomes fairly symmetrical and, at first glance, it could be considered as a single band, more intense and centred at 430 nm, as was the case with irradiation at 0° in the spectrum at normal incidence. When the sample is left to rest, the absorption continues to increase, but undergoes a small hypsochromic shift. This effect increases progressively with the time elapsed since irradiation in the stored films. Similar results were obtained using other irradiation wavelengths, such as 390 and 500 nm, however, with 440 nm the phenomenon is fast and leads to highly reproducible results.

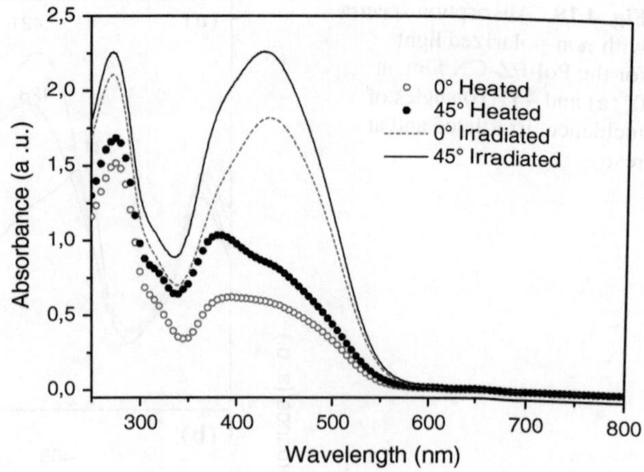

Fig. 4.19 Absorption spectra at normal and oblique incidence for a Pol-PZ-CN film that has been heated to maximum Δn and afterwards irradiated

The increased absorption is due to the sample disaggregating, since, as mentioned, the isolated azobenzenes have higher oscillator strength than the aggregated ones.

If the same irradiation process is applied to a film with the maximum out-of-plane birefringence achieved by heating above the mesophase temperature, the absorption spectrum of Fig. 4.19 is obtained for unpolarized light at normal and oblique incidence.

As can be seen in this figure, the absorption increases substantially, because the amount of aggregates in a heated film is very large, and they have partly evolved to isolated azobenzenes. In the bands of the irradiated film some shoulders are observed at about 390 nm, corresponding to aggregates still present in the film, which can be eliminated by subsequent irradiation.

Moreover, it is noteworthy that the re-dissolution and evaporation of films that have been irradiated or heated above the mesophase temperature leads to an absorption spectrum very similar to the original one, so that the initial "aggregation state" of the films is a result of the film formation mechanism.

Measurement of the refractive indices of the films after the mentioned irradiation times ($\cong$ 1 min) does not reveal a significant change in Δn, which remains practically the same as in freshly prepared films.

Once the possibility of modifying the aggregation state of the film by light has been recognized, it is interesting to explore the possibility of controlling or modifying the alignment of the molecules in the film also by irradiation. As discussed above, the starting samples have a strong tendency to the axial out-of-plane orientation. We considered whether it is possible to optically induce high levels of in-plane Δn and whether these values can be increased by thermal treatments (thermotropic effect). In order to achieve high anisotropy in the plane, films were irradiated with horizontally polarized light at normal incidence and also as oblique as possible (about 70°). The axes of the film and the direction of irradiation are defined in Fig. 4.20.

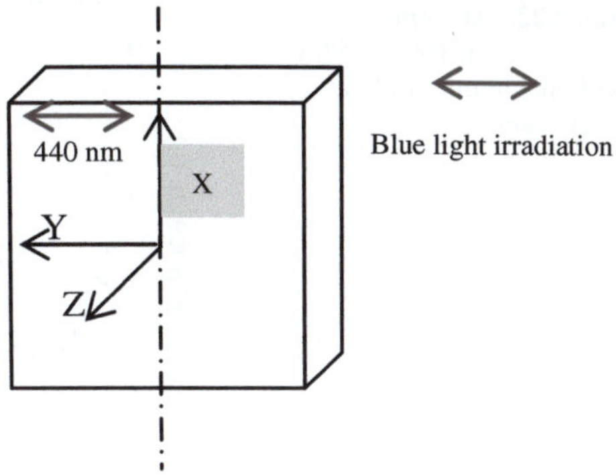

Fig. 4.20 Scheme showing the irradiation direction of the 440 nm light with respect to the film axes

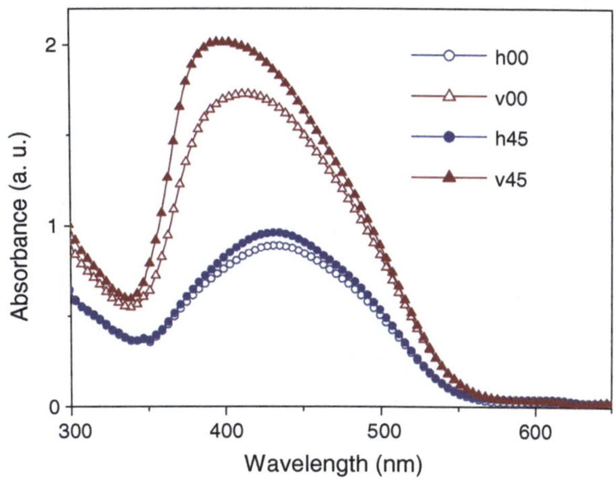

Fig. 4.21 Absorption spectra of Pol-PZ-CN films irradiated with horizontally polarized light

After irradiation, the spectra at normal (00) and oblique (45) incidence were obtained for the two polarization directions of the incident light, as shown in Fig. 4.21. The horizontal polarization direction (h) matches the *Y*-axis, and the vertical one (v) the *X*-axis of Fig. 4.20.

As can be seen in Fig. 4.21, a significant dichroism has been achieved in the plane through this process, although a homeotropic trend can still be observed, since the absorption at 45° is higher than at normal incidence.

The refractive index measurement of irradiated films in the situation corresponding to Fig. 4.21 resulted in $n_x = 1.74$, $n_y = 1.64$ and $n_z = 1.72$. It is observed that the in-plane index, in the *X* direction, perpendicular to the irradiation, is slightly higher than that measured in the *Z* direction, perpendicular to the plane. In this situation, if the film is heated above the mesophase transition temperature and maintained for 30 s at 120 °C, the azobenzenes align in the *X* direction (maximum index), owing to the thermotropic effect, and the following refractive indices are obtained: $n_x = 1.88$, $n_y = 1.55$, and $n_z = 1.60$. The achieved birefringence is comparable to that obtained out-of-plane by exclusively thermal treatments $\Delta n = 0.35$). The absorption spectrum corresponding to a film subjected to this

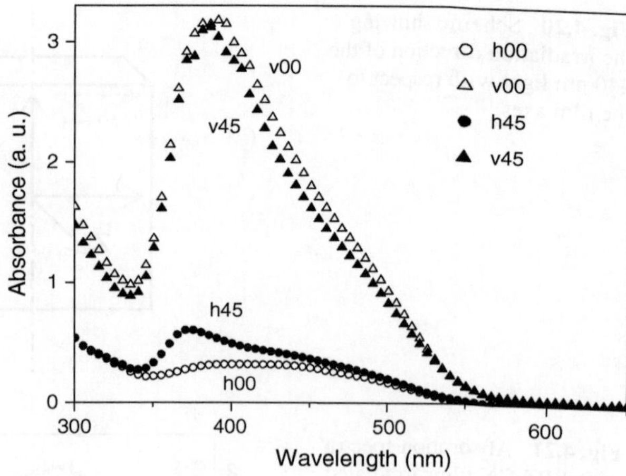

Fig. 4.22 Absorption spectra of Pol-PZ-CN films with maximum in-plane birefringence

treatment (irradiation + 30 s at 120 °C) and leading to high in-plane anisotropy is shown in Fig. 4.22.

In view of the figure, it can be checked that the anisotropy obtained in the measurement of refractive indices corresponds to a significant in-plane dichroism. Likewise, the tendency to form aggregates is evidenced in these films when its temperature rises above the mesophase, since there is a clear hypsochromic shift of the absorption maximum, related to the increase of the contribution of the H aggregate's band at 380–390 nm.

It is also noticeable that a large enough value of in-plane Δn ($n_x > n_z$) is necessary in order to produce the in-plane thermotropic effect.

4.1.2 Nonlinear Response

After studying the effects of light and temperature on the arrangement and the degree of order acquired by the azo chromophores, we studied the nonlinear response of these systems. Firstly, an investigation of the macroscopic response of the polymers by means of SHG was proposed, relating it to the molecular hyperpolarizability values, in order to analyse the degree of polar order established after the poling process. Secondly, the films were subjected to different optical and thermal treatments, whose effect on the nonlinear coefficients values was also analysed.

4.1.2.1 As-Prepared Films

Two groups of films of Pol-O-CN and Pol-PZ-CN were prepared by evaporation of CH_2Cl_2 on ITO glass substrates. The first group was characterized by a thickness of about 1.2 μm and the second, of about 0.6 μm.

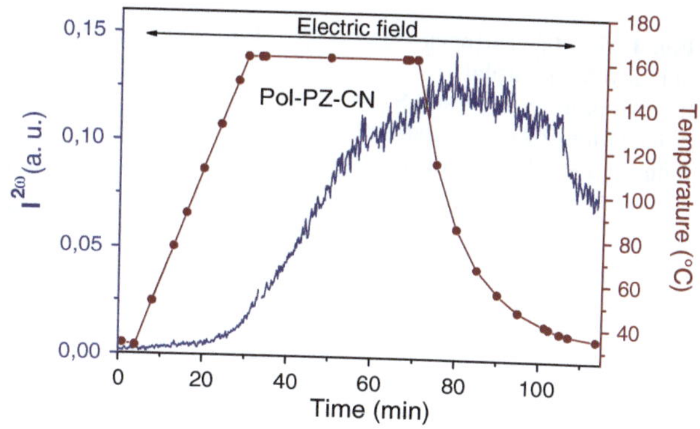

Fig. 4.23 In situ monitoring of the second harmonic intensity generated during the thermal poling of a freshly prepared Pol-PZ-CN film. The P^{ω} polarized fundamental light strikes the film at 40°

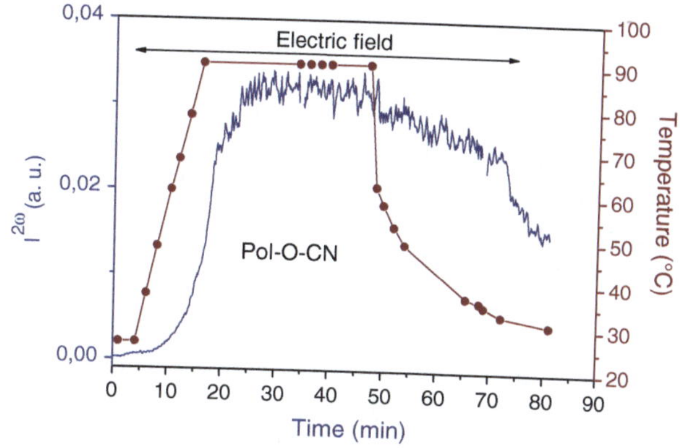

Fig. 4.24 In situ monitoring of the second harmonic intensity generated at an angle of incidence of 40° during the thermal poling of a freshly prepared Pol-O-CN film

In a common thermal poling process, the electric field is switched on at room temperature, heating the sample at a rate of 5 °C/min up to the poling temperature T_p, at which temperature the film is held for 40 min. One of the criteria used to establish the optimal value of T_p is based on monitoring the second harmonic signal while gradually increasing the temperature of the film. After a certain temperature value, the generated intensity decreases as a result of an increase in conductivity, sample deterioration, etc. Usually, we set T_p about 10–15 °C below that level, since the film will be kept for about half an hour at that temperature. In the case of Pol-PZ-CN films that were thicker than 1 μm and had not been exposed to previous optical treatments, the temperature was raised to 170 °C, and a clear decrease of the signal during the process was not observed. However, when working with thinner films, and especially if they had been previously irradiated with blue light, raising the temperature led to a signal decrease.

Figure 4.23 shows the orientation by thermal poling of a film of this material (up to 160 °C). The evolution of the second harmonic signal generated by the film simultaneously with the application of the field can be observed. In these measurements, horizontally polarized excitation light is used, because the generated intensity is greater.

The poling profile for Pol-O-CN films is shown in Fig. 4.24 for comparison. As can be seen, the initial slope is sharper in this case, that is, the intensity of the

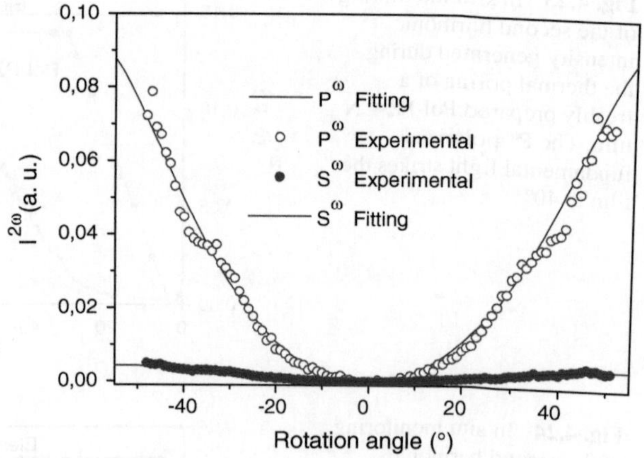

Fig. 4.25 Maker fringes obtained by the usual thermal poling for P^{ω} and S^{ω} polarizations of the excitation light for a Pol-PZ-CN film

second harmonic signal rises faster. This is connected with the lower incidence of the aggregation in the case of Pol-O-CN, and will be discussed later. The two profiles show a drop when the electric field is removed that is associated with the neutralizing of the trapped charges in the film surface during poling and with different relaxation processes [12–14].

Once the films had been oriented according to the procedures described, the characterization of their nonlinear response was performed. In order to do this, we measured the second harmonic signal generated by the films as a function of the angle of incidence of the excitation light. As described in Chap. 2, the measurements were usually carried out for two orthogonal polarizations of the incident light, P^{ω} and S^{ω}. In Fig. 4.25, a typical set of results, which gives rise to so-called Maker fringes, is shown. These experimental curves are fitted to the expression 1.26 to obtain the nonlinear coefficients d_{31} and d_{33}: the former is obtained by fitting the curve measured with S^{ω} polarization, while the fitting of the curve obtained with P^{ω} polarization, which depends on both coefficients, leads to d_{33} once d_{31} is known.

When it comes to giving a value of the nonlinear coefficients of the systems concerned, it should be recalled that the polar order induced by applying an electric field in polymers is not stable in general, resulting in a decrease of the second harmonic intensity generated by the films as time passes, starting from the poling process.

It is therefore necessary to measure the response of the films after different time intervals after the polar orientation process, in order to study the evolution of the nonlinear coefficients and to draw conclusions about the stability of the nonlinear response. The evolution of the effective nonlinear coefficient versus time elapsed since poling is shown in Fig. 4.26, calculated from the experimental values of d_{33} and d_{31} by using the expression

$$d_{eff} = d_{31}\cos\left(\theta'_{2\omega} - \gamma_{2\omega}\right)\sin 2\left(\theta'_{\omega} - \gamma_{\omega}\right) + \sin\left(\theta'_{2\omega} - \gamma_{2\omega}\right)$$
$$\left[d_{31}\cos^2\left(\theta'_{\omega} - \gamma_{\omega}\right) + d_{33}\sin^2\left(\theta'_{\omega} - \gamma_{\omega}\right)\right]$$

for an incidence angle of 40°.

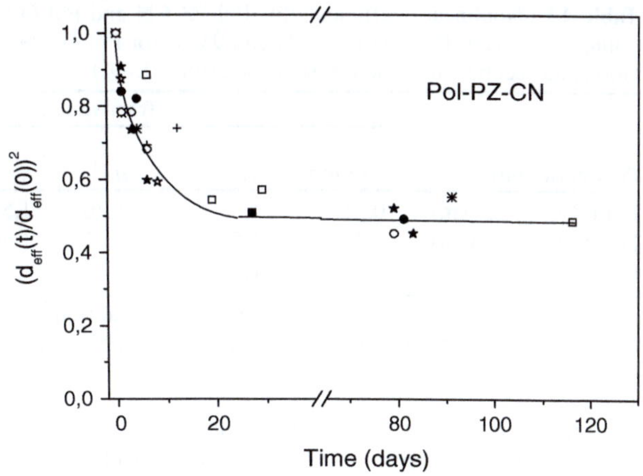

Fig. 4.26 Time evolution of the NLO response for several Pol-PZ-CN films normalized to the value measured approximately 2 h after the electric field was turned off

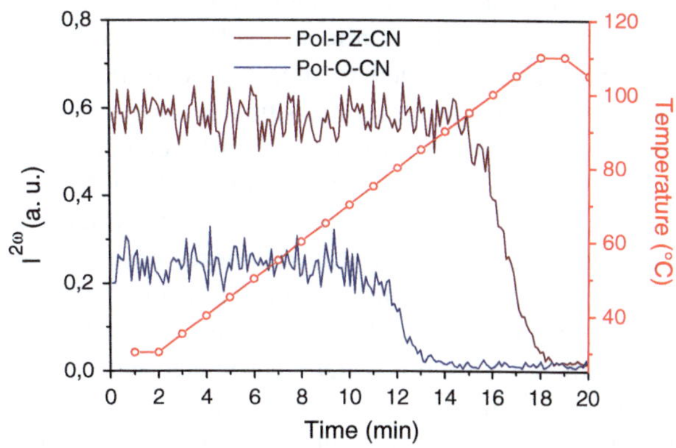

Fig. 4.27 Second harmonic intensity evolution with temperature for the Pol-O-CN and Pol-PZ-CN films

This coefficient is proportional to the detected intensity for P^{ω} excitation. In the figure is shown the square of the coefficient normalized to the value corresponding to the recent poling (2 h). In the graph the results obtained for several films have been considered.

As can be seen, after 3 weeks a stable value is achieved that remains essentially constant for several weeks after poling. By contrast, in the case of Pol-O-CN, the signal decreases to 40 % in just one week.

Along with temporal stability, it is interesting to check the thermal stability of the induced polar order of the two polymers. With this goal, we measured the second harmonic signal with 40° incident angle of the excitation light, while heating the samples at 5 °C/min. As shown in Fig. 4.27, after a certain time interval in which the signal remains stable, it begins to drop at a given temperature, until it finally disappears. As expected, in the Pol-PZ-CN polymer the second harmonic signal is maintained at a higher temperature than in Pol-O-CN. For the former, the deletion of the signal starts between 90 and 100 °C, while for the latter, the signal

Table 4.5 Nonlinear coefficients for Pol-PZ-CN and Pol-O-CN films two weeks after thermal poling ($T_p = 120$ °C for Pol-PZ-CN and 95 °C for Pol-O-CN), and refractive indices at the excitation and second harmonic frequencies (error ±15 %)

	Refractive indices					d_{ij} [pm/V]	
Polymeric film	Δn at 633 nm	$n_x = n_y^{\omega}$	n_z^{ω}	$n_x = n_y^{2\omega}$	$n_z^{2\omega}$	d_{31}	d_{33}
Pol-PZ-CN (1.2 μm)	0.21	1.55	1.70	1.56	1.73	1.0	8.0
Pol-PZ-CN (0.6 μm)	0.22	1.55	1.72	1.56	1.74	1.3	10
Pol-O-CN (1.2 μm)	0.18	1.56	1.70	1.57	1.73	0.5	3.5

The superscript ω refers to the excitation fundamental frequency in SHG measurements: $\lambda = 1907$ nm. 2ω refers to the second harmonic of the excitation frequency: $\lambda = 954$ nm

is maintained up to 70–80 °C. This result is consistent with the glass transition temperature, which is 20 °C lower in Pol-O-CN. Upon reaching T_g, the system starts showing fluidity, i.e., the polymer chains acquire a certain mobility, and thus it loses the achieved polar orientation.

The values of the nonlinear coefficients and refractive indices that are shown in Table 4.5 correspond to measurements performed two weeks after they were oriented, when the harmonic signal evolution is very small. These values reflect the average of several films with similar characteristics under analogous processes.

The values of the coefficients for Pol-PZ-CN, as can be deduced from the table, are somewhat higher for thin films than for thicker ones, which may be due to the fact that in thin films the chromophores are more accessible to be oriented by the electric field, which would be more uniform.

A first fact that can be highlighted is that the obtained d_{33}/d_{31} ratio is clearly higher than 3, which is usually associated with a high polar order (higher than that achieved in amorphous polymers with the same field conditions as those used in this work). This result, which is related to the fact that the polymers are liquid crystals reaching very anisotropic conditions, will be discussed in more detail when we analyse the NLO response of pre-irradiated films.

In view of the nonlinear coefficients, the response observed for Pol-PZ-CN is greater than that for Pol-O-CN. It is an expected result, given the previously analysed molecular response, where the $\mu\beta$ value was much lower for mon-O-CN than for the chromophore with piperazine. However, if we compare the values quantitatively, they do not appear to meet the expected relationship. We will see this more clearly if we compare the d_{33} values obtained for freshly oriented samples, since the different polar order relaxation processes, much more evident in Pol-O-CN, can obscure the phenomenon. In the case of Pol-PZ-CN the coefficient is $d_{33} = 11$ pm/V and for Pol-O-CN the result is $d_{33} = 6.5$ pm/V. The ratio between values of $\mu\beta$ for the two polymers is about 3.8 (see Table 3.1), however, at a macroscopic level, we see that the ratio of the experimentally obtained d_{33} coefficients is less than 2, well below the predicted value at the molecular level.

In order to explain this difference, we will use the estimate of the d_{33} value predicted by the model for oriented dipoles (oriented rigid gas), as discussed in Chap. 1.

We will calculate these values by using expression 1.43: $d_{33} = 1/2\left(N\beta^*_{zzz}\right.$ $\left.\langle\cos^3\theta\rangle\right)$, where β^*_{zzz} is the microscopic hyperpolarizability incorporating local field factors and $<\cos^3\theta>$ the orientational average of $\cos^3\theta$, which is related to the polar order parameters (see Appendix C) and to the angle θ between the chromophore and the direction of the poling.

We can compare the estimated maximum value of d_{33} that they would achieve theoretically, knowing the dipole moment of the two molecules (6 D for the chromophore with the O donor group, and approximately 9 D [15] for the one with piperazine; 1 D (debye) $\approx 3.34 \times 10^{-30}$ C m). The local field factors defined in Chap. 1 are estimated to be 4.6 and we can calculate the number of molecules per unit volume in the film. We take 1.1 g/cm^3 as an approximate density for the polymers, since that of PMMA is 1.18 g/cm^3 and there is a published value of 0.98 g/cm^3 for a similar system PMMA/DANS [16], where DANS is 4-dimethylamino-4′-nitro-stilbene. With these values, we would have maximal d_{33} coefficients (assuming a perfect polar order, that is, $<\cos^3\theta> = 1$) of 140 and 30 pm/V for Pol-PZ-CN and Pol-O-CN, respectively.

In amorphous polymers oriented by a field, where all the induced anisotropy is associated with the polar order, it is usual to estimate this polar order from absorption or refractive index measurements, which actually provide the value of P_2, the axial order parameter. In our case, the anisotropy value measured in the area of the film oriented by the field is similar for both polymers ($\Delta n \approx 0.2$), which could be associated with similar degrees of polar order; then the $<\cos^3\theta>$term would be equivalent in the two cases. Therefore, the ratio that should be obtained between the d_{33} values of the two polymers would be, in the case of the maximal values, larger than that obtained experimentally.

This seeming disagreement disappears if one considers that not only the chromophores that are polarly oriented by the electric field contribute to the axial order, there is also a contribution from the chromophores that tend to be placed perpendicularly to the substrate, which can form nonpolar aggregates, which cannot be oriented by a field. When studying the effect of thermal treatment on the films, we have seen that heating results in a very important out-of-plane optical anisotropy. We propose, therefore, that a large part of the measured birefringence in the area where the electric field is applied at high temperature is actually related to the nonpolar axial order, especially in the case of Pol-PZ-CN.

One of the results supporting this hypothesis is the value of the birefringence measured several weeks after poling in samples subjected to polar orientation processes. As discussed in the text, the value this birefringence achieves in the area oriented by the field is around 0.2 (when the temperature $T_p = 120$ °C). Several weeks later, when the second harmonic signal has decreased considerably (see Fig. 4.26), Δn is virtually unchanged. If the birefringence were associated mainly with the polar order, a corresponding decrease should be observed. Something similar happens when the second harmonic signal is thermally erased, as shown in Fig. 4.27. During this process, the films are kept at high temperature just long enough for the harmonic signal to become zero. After this process, the measurement of the indices shows that the optical anisotropy has not been completely erased.

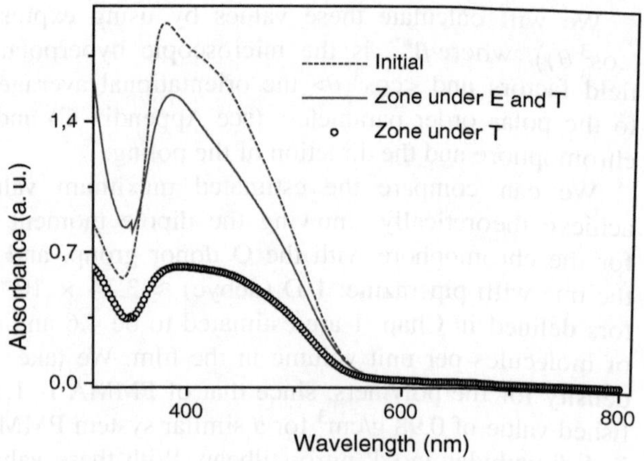

Fig. 4.28 Absorption spectra for the freshly prepared film, one subjected to $T > T_{mesophase}$, and one subjected to $T > T_{mesophase}$ plus electric field

Having ascertained that not all the order induced by applying an electric field at high temperature (represented by the Δn value in the table) is polar, the question of how far the orientation of the chromophores is affected by the electric field arose. Thus, we measured the optical anisotropy of freshly prepared films subjected to a poling process, comparing what happens in the central area, which was more strongly affected by the field, and in the surrounding areas. The average value of the birefringence at 633 nm obtained for Pol-PZ-CN films oriented this way ranges from 0.15 to 0.2, rather less than that measured in the areas that were not affected by the electric field. These areas only reached temperatures above the mesophase temperature for about 40 min, analogous to Fig. 4.15, and they reached birefringence values exceeding 0.3. Furthermore, they exhibit a lower absorption than the area where the electric field was very intense, as shown in Fig. 4.28.

Moreover, the NLO response was measured for films oriented by thermal poling, but in which maximal out-of-plane birefringence had been previously induced by heating. The NLO coefficients obtained were clearly lower than in the case of freshly prepared films.

These results suggest that the electric field is not able to overcome the chromophore's tendency to aggregate in Pol-PZ-CN during the poling process, although it is disturbed. There is competition between the electric field orienting the molecular dipoles and the intermolecular interactions themselves. By increasing the poling temperature, the formation of nonpolar aggregates (which cannot be oriented by the field) is favoured. The relatively low nonlinear response of this polymer is associated, therefore, with the aggregation already present in freshly prepared films and favoured at high temperature. The application of an electric field does not disaggregate it or prevent its formation.

With respect to Pol-O-CN, we have seen that the experimental value of d_{33} is closer to the maximum estimated for Pol-PZ-CN. This can be explained considering that in the polymer with piperazine the larger dipole moment favours the chromophore–chromophore interactions, and therefore the formation of nonpolar aggregates, which do not contribute to the nonlinear response. In the case of Pol-O-CN, the lower dipole moment of the azo chromophores causes interactions that

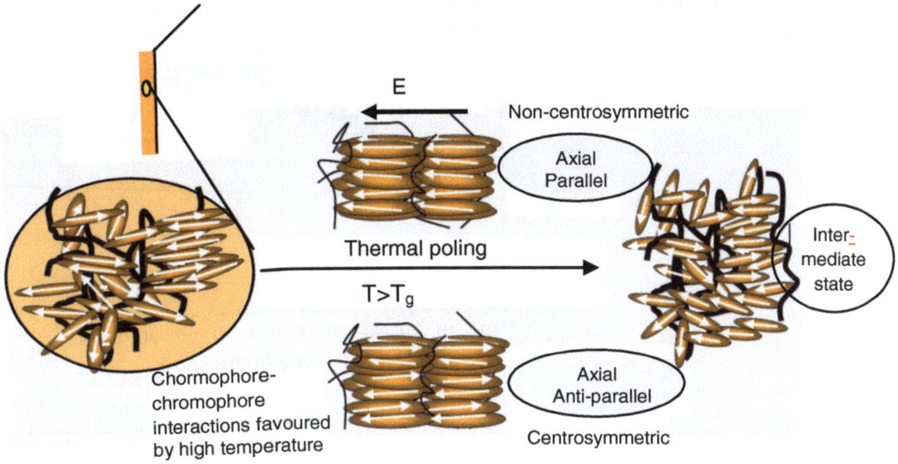

Fig. 4.29 Ideal representation of the movement of the side chains in PCL films during the thermal poling process

are not so strong and so neither is the tendency to form stable aggregates, making it easier for the electric field to align a greater number of chromophores.

So, we can draw the conclusion that the electric field–induced orientation and that associated with the thermotropic effect compete for the alignment of the chromophores in a parallel, and consequently non-centrosymmetric way, in the first case, and in an antiparallel and centrosymmetric manner in the second case. This phenomenon is illustrated in. Figure 4.29. In the figure the chromophores in the freshly prepared film are depicted with a nonpolar axial trend (homeotropic) order. When the film is subjected to a corona discharge and high temperature an intermediate state is achieved in which part of the axial orientation is polar, and other part is antiparallel.

4.1.2.2 Films Treated with Blue Light

I. Unpolarized light effect

a. *Previous irradiation*

As has already been explained, the aggregation state of the films influences the final polar organization achieved, so it is desirable to start from a less aggregated state than the one shown by films prepared by evaporation. Therefore, the thermal poling process was applied to films pre-irradiated with unpolarized light of 440 nm wavelength. In this previous process, the aim is to access as many aggregates as possible. Since, as we have seen, the aggregates tend to be positioned homeotropically, the light was focused on the film at angles up to 70°, as our goal was to break as many possible.

In Fig. 4.30 the thermal poling profile shown by the pre-irradiated Pol-PZ-CN films of thickness $\cong$ 1.2 μm is represented. As shown in this figure, the

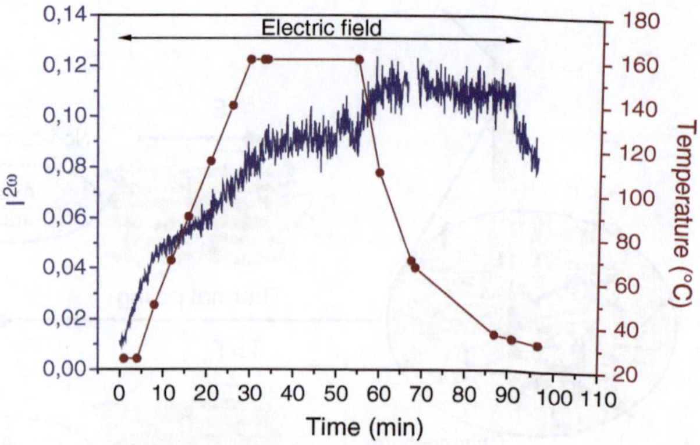

Fig. 4.30 Thermal poling of the pre-irradiated Pol-PZ-CN film

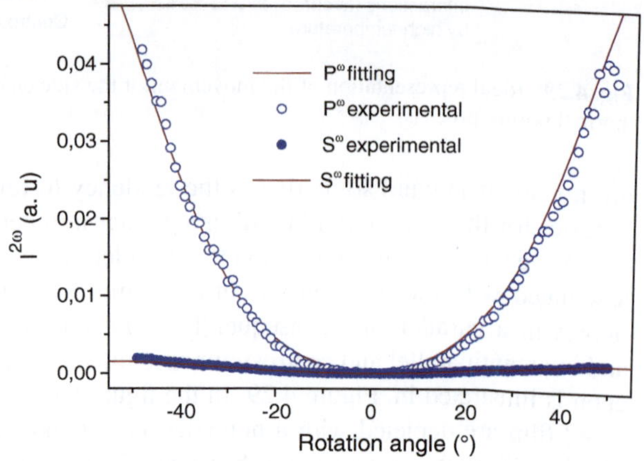

Fig. 4.31 Maker fringes for P^{ω} and S^{ω} polarizations of the excitation light for Pol-PZ-CN films

initial increase of the signal is faster than in untreated samples. This is because the irradiation produces *cis* isomers, breaking aggregates, and the global mobility of the material is larger [17–19], owing to the increased fr0065e volume. Also, slope changes are seen during heating, which could be associated with the different mentioned axial orientation mechanisms. As can be observed in Fig. 4.30, as the temperature increases, the rapid initial polar orientation slows down, since the aggregation and antiparallel orientation are thermally favoured, even though there is a greater mobility than at low temperature. The initial slope of this graph is similar to the case of Pol-O-CN, where, as mentioned, aggregation does not take place so noticeably.

The nonlinear response of the pre-irradiated films was characterized also by SHG measurements. An example of Maker fringes obtained with pre-irradiation, measured for two polarizations of the excitation light, parallel P^{ω} and perpendicular S^{ω}, is shown in Fig. 4.31.

The stable d_{ij} coefficients values, measured two weeks after the poling process, and the refractive indices for the fundamental wave and the second harmonic are shown in Table 4.6.

Table 4.6 Refractive index values at 1.9 μm and 953 nm (ω and 2ω) and stable NLO coefficient values for pre-irradiated Pol-PZ-CN films with two different thicknesses

Polymeric film	Refractive indices				d_{ij} [pm/V]	
	$n_{x=y}^{\omega}$	n_{z}^{ω}	$n_{x=y}^{2\omega}$	$n_{z}^{2\omega}$	d_{31}	d_{33}
Pol-PZ-CN (1.2 μm, 160 °C)	1.53	1.73	1.55	1.76	1.0	12
Pol-PZ-CN (0.6 μm, 120 °C)	1.53	1.78	1.54	1.80	1.1	14

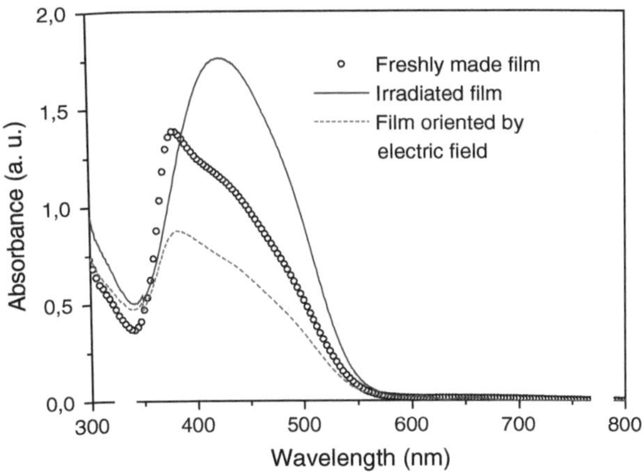

Fig. 4.32 Absorption spectrum evolution during the poling process with pre-irradiation

As deduced from the data shown in the table, the nonlinear response is larger than in untreated films. The optimal result was found in thin films at 120 °C. The d_{ij} coefficients obtained for freshly oriented films in such circumstances were $d_{33} = 17$ pm/V and $d_{31} = 1.3$ pm/V, whereas, for films of equal thickness but untreated with light they were $d_{33} = 13$ pm/V and $d_{31} = 1.7$ pm/V under the same conditions.

The measurement of the refractive indices of these films showed that the birefringence in the electric field-oriented area is greater than in the samples without pre-irradiation, which is related to the increased mobility of the chromophores. The shape of the absorption band obtained after poling is similar to the initial one (before irradiation) but less intense, showing a clear peak due to the aggregates centred at about 390 nm, as shown in Fig. 4.32. It seems, therefore, that the light disrupts aggregates and groups of oriented molecules, allowing a greater change in axial direction by using electric field and temperature.

The results in Table 4.6 show that, by using pre-irradiation, higher values for d_{33} and for the ratio d_{33}/d_{31} are obtained. Part of the observed increase in the NLO response is associated with the fact that disaggregation increases the number of free chromophores, which are now accessible to be oriented by the field. Regarding the coefficient ratio, we will bear in mind, as explained in Chap. 1, that in a polymer oriented by electric field the chromophores are distributed around the electric field director axis. For initially isotropic media, the d_{33}/d_{31} ratio is 3 according to the mentioned thermodynamic model, where the induced order is due to the orientation achieved by using low fields [20]. For high field values, d_{33} approaches its saturation value and yet d_{31} approaches zero, so d_{33}/d_{31} ratios greater than 3 have been associated with a higher polar order caused by strong electric fields [21, 22] or interactions between chromophores that favour highly axially oriented states, as occurs in the case of side-chain LCPs, according to different researchers [23–25]. In order to establish whether the applied field is very large and therefore if it is questionable to consider the low-field approximation, films of a mixture of amorphous PMMA with a small amount of DR1 were prepared and a conventional thermal poling was applied in the same conditions as for Pol-PZ-CN.

The fitting of the fringes measured for PMMA/DR1 film resulted in a ratio $d_{33}/d_{31} \cong 3$. Thus, it seems that the applied electric field is not very strong, and the high values for this ratio found in our polymers are associated with the LC nature of the materials in question, which are highly axially oriented, as can be deduced from the registered values of the birefringence. Since the birefringence values are higher in the pre-irradiated samples, the ratio is also higher.

b. In situ *irradiation (photoassisted poling)*

By using the photoadressability of the azobenzene it is possible to perform a polar orientation process of the azo polymers at a low temperature far from its T_g by so-called photoassisted poling, to avoid possible degradation. Photoassisted poling was tested for two different wavelengths, 406 and 488 nm. At equal power, there were no noteworthy differences, so we show the results obtained with an Ar laser emitting at 488 nm. The laser light is brought to the NLO equipment through an optical fibre, which offers the possibility of obtaining depolarized light. The study carried out using different light powers revealed that there is an optimum power ($\approx$ 100 mW/cm^2) above which its improving effect on the orientation with the field decreases. This is probably because the light induced such mobility that it caused a disorder effect instead of assisting the electric field to achieve a polar orientation. Therefore, the power at which the phenomenon was clearly observed was used, and the process was performed in two films of equal thickness, one freshly prepared and the other one previously irradiated, as was done for thermal poling, in order to see the effect of "disaggregation" prior to this process. Excitation light polarized parallel to the plane of incidence, P^{ω}, was employed to monitor the signal.

The previous process of disaggregation seems to influence the initial part of the orientation process, as seen by comparing Fig. 4.33a and b. In Fig. 4.33b a gradual increase in the signal with the electric field is observed, owing to the increased mobility in the pre-irradiated material because of the disaggregation. At the moment at which the film is irradiated with light, we observe a decrease of the signal in both cases, larger in the pre-irradiated one, which is due to the disorientation of some of the previously polarly oriented chromophores.

In the case of the pre-irradiated film, the slope during irradiation does not vary noticeably with respect to the previous phase of the exclusive electric field orientation. However, in the freshly prepared film a clear increase of the slope can be seen. This suggests that the mobility gained by the material during in

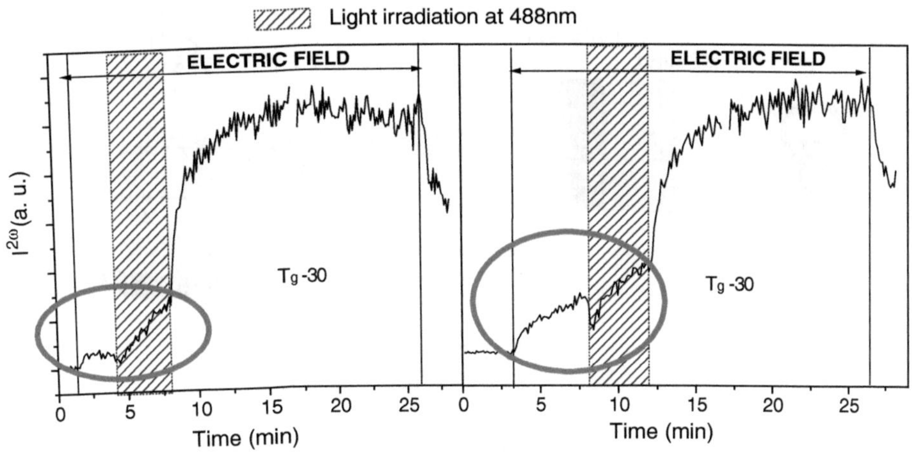

Fig. 4.33 Photoassisted poling of Pol-PZ-CN films. **a** Freshly prepared film. **b** Pre-irradiated film

situ radiation is due to the disaggregation. In both cases, when the irradiation is stopped, there is a clear increase of the slope, because of the *cis* isomers generated in the previous phase, whose presence increases the free volume in the material, and thus facilitates the final orientation of the *trans* isomers with the field.

In the films oriented this way, the coefficients calculated from Maker fringe measurements were $d_{33} = 7$ pm/V and $d_{31} = 1$ pm/V. These values are less stable than those achieved thermally. This is because although the light increases the molecular mobility of the dipoles, in the main polymer chain the ordering is hardly induced below T_g [26–28]. By contrast, in thermal poling, local rearrangements occur in the chain, around the oriented dipoles, resulting in a greater stability of the induced polar order.

In order to optimize the NL response an orientation protocol by means of thermal poling was proposed, which includes in situ irradiation with 488 nm wavelength light during the first stage at low temperature. The improvement that light introduces is to be exploited with the stability that the high temperature poling provides. The values of the coefficients and the d_{33}/d_{31} ratio thus obtained were the highest found for Pol-PZ-CN: $d_{33} = 17$ pm/V and $d_{31} = 1$ pm/V. The Maker fringe recording measured 15 days after the orientation showed very good stability, and after 6 months a signal that resulted in very high coefficients, $d_{33} = 12$ pm/V and $d_{31} = 0.7$ pm/V, still remained.

II. Linearly polarized light effect

a. *Previous irradiation*

In the section dealing with the optical study, it has been proved possible to overcome the homeotropic tendency of the Pol-PZ-CN material by irradiation

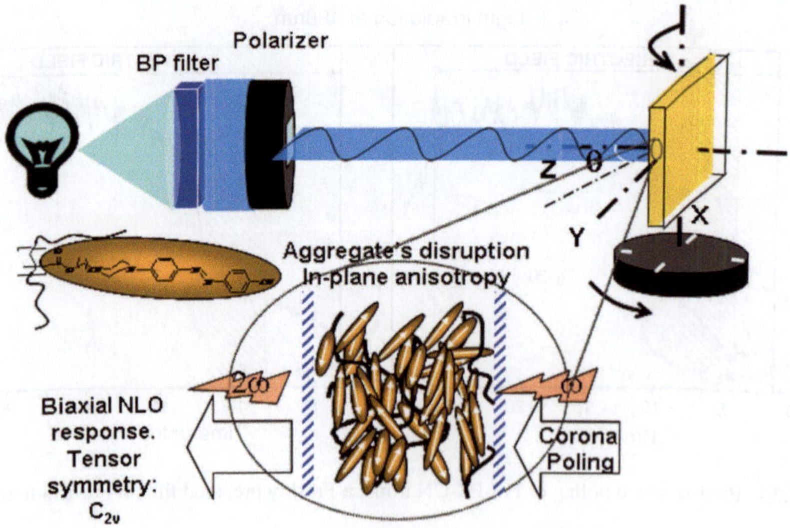

Fig. 4.34 Schematic representation of the two-step polar orientation process (in-plane anisotropy generation by means of irradiation and thermal poling). *BP filter*: Bandpass filter

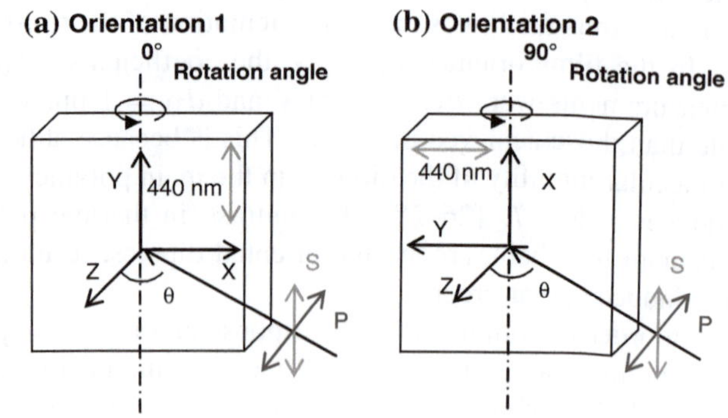

Fig. 4.35 Scheme of the configurations of the films for pre-irradiation and experimental measurements of Maker fringes

processes that control the arrangement of the molecules in the film. We propose here the application of thermal poling to pre-irradiated samples with in-plane anisotropy, to assess the possibility of forcing different symmetries than those usually found in the oriented polymers. Thus, after applying this process, a material with biaxial NLO response is obtained, as outlined in Fig. 4.34. It is a two-step process, in which, firstly, the film is irradiated with linearly polarized light using orientation 2 of Fig. 4.35.

The films were irradiated at oblique angles of incidence in both directions to minimize any possible out-of-plane anisotropy and to access the largest number of aggregates. Thus, values were obtained in which one of the in-plane indices, in the *X* direction, is greater than that measured in the direction perpendicular to the film,

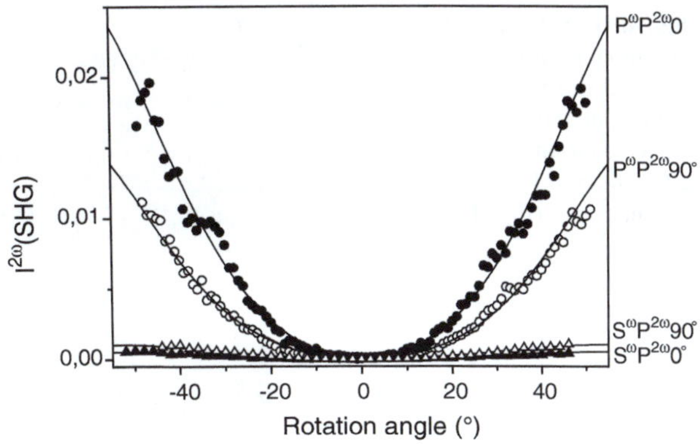

Fig. 4.36 Maker fringes for a Pol-PZ-CN film with in-plane anisotropy, for two configurations, 0° and 90°, and two polarizations of the excitation light, P^{ω} and S^{ω}

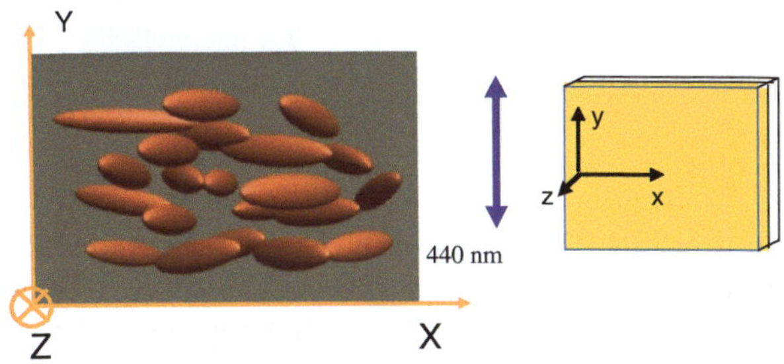

Fig. 4.37 Orientation scheme of the chromophores in the films irradiated with polarized light

Z. As discussed in the optical study section, this condition is necessary to make the biaxial symmetry prevail over the axial trend (with the axes perpendicular to the plane), which is forced by the thermotropic effect.

Below, the standard orientation process by applying a field at high temperature (120 °C) is applied to the film irradiated under the conditions described above. Under these conditions the Maker fringe measurement show several different features from those described so far. If we analyse the fringes measured with either of the two polarizations used, P^{ω} or S^{ω}, we find that the intensity of the fringes depends on the orientation of the film, as it rotates around an axis that is perpendicular to the plane of the film itself. That is, if we consider orientations 1 and 2 of Fig. 4.35, the measured intensity is clearly different. Moreover, the ratio of the intensities of the fringes measured with different polarization $I(P^{\omega})/I(S^{\omega})$, also depends on the orientation of the film. Specifically, in orientation 1, the P^{ω} intensity is larger than in orientation 2, while in the fringes measured with S^{ω} polarization we have the opposite situation, with a larger intensity in orientation 2. An example of these results is shown in Fig. 4.36.

This lack of axial symmetry is related to the in-plane optical anisotropy described at the beginning of this chapter. That is, the optically induced biaxiality has been preserved during the poling process and results in a nonlinear response that is also biaxial. Figure 4.37 shows an idealized picture depicting the situation

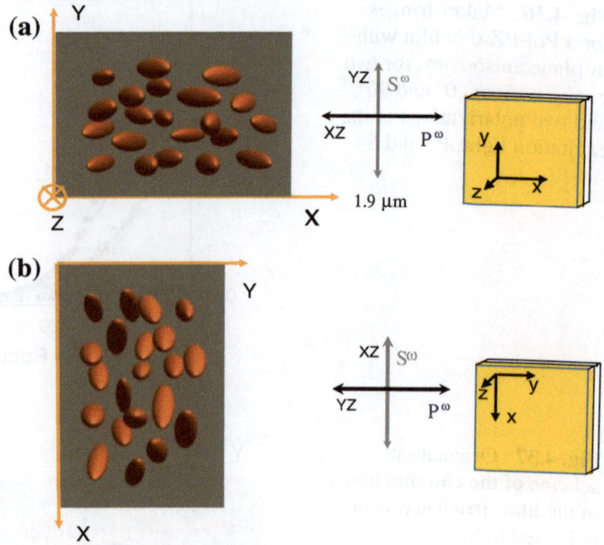

Fig. 4.38 Scheme of the chromphore orientation in the films irradiated with polarized light and subjected to thermal poling. **a** Orientation 1 (Fig. 4.35). **b** Orientation 2 (Fig. 4.35)

in which the chromophores are preferably in the *ZX* plane, once the film has been irradiated with light with vertical polarization (*Y* direction) during the first step of the process. To simplify the drawing and make it clearer, the polymer chain to which the chromophores are joined is not represented. As can be seen, the direction of the chromophores is preferably homeotropic (*Z* direction in the drawing), since the starting films show out-of-plane birefringence.

The second step of the orientation, as was discussed, consisted in the application of a thermal poling. In Fig. 4.38a and b we can see the orientation that the chromophores acquire in an ideal model after the treatment with electric field and high temperature, starting from the situation of Fig. 4.37.

In this measurement configuration, that is, that in Fig. 4.38a (orientation 2), the fringes obtained with P^{ω} polarization will show less intensity, since the orientation of the chromophores has a smaller component in the *YZ* plane direction, which is the *XZ* of orientation 1. Moreover, the fringes with S^{ω} light will present greater intensity in orientation 2 than in 1, as the orientation has a larger component in *XZ* than in *YZ*.

The in-plane anisotropy influence on the NLO properties of polymers has been explored by several authors. However, the orientation has been obtained mainly using mechanical methods. For example, the control of the refractive indices in the three directions for polymers by means of drawing or tensile strength was described [29]. In this case, moreover, the SHG in the phase-matching configuration was studied by using the achieved in-plane birefringence for these polymeric systems [30]. In this way, the highest possible efficiency of conversion of light into the second harmonic is achieved, which increases the interest in these materials for applications in nonlinear waveguides. Yet, no reference can be found to this phenomenon in polymers with photoinduced in-plane anisotropy.

Table 4.7 Values of the three refractive indices and nonlinear coefficients for the pre-irradiated Pol-PZ-CN films

	Refractive indices			d_{ij} [pm/V]		
Pol-PZ-CN (0.6 μm)	n_x	n_y	n_z	d_{31}	d_{24}	d_{33}
Pre-irradiated pol. light	1.68	1.57	1.80	1.4	0.9	11
Pre-irradiated pol. light + heating	1.87	1.55	1.61	1.5	0.5	5

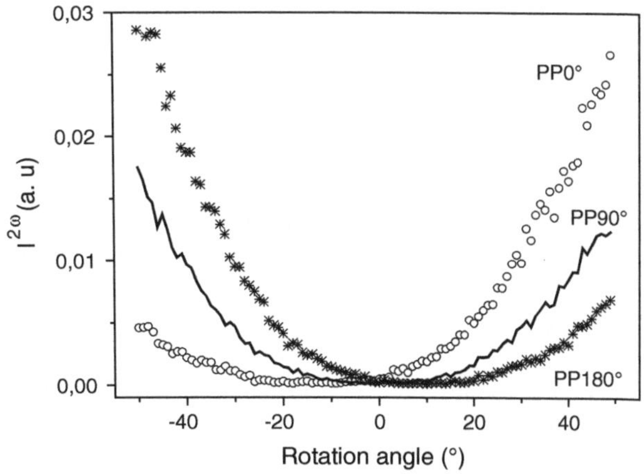

Fig. 4.39 Maker fringes for a Pol-PZ-CN film irradiated at 45° with unpolarized light

The nonlinear response of these films was quantified from the curve fittings in Fig. 4.36, from which we obtained the values of the three independent nonlinear coefficients, d_{33}, d_{24} and d_{31}, which are compiled in Table 4.7, together with the refractive index values measured at 633 nm.

In Table 4.7 the values of the NLO coefficients and the refractive indices for a film oriented by field but with maximum in-plane anisotropy are also listed. This anisotropy is achieved by irradiation with linearly polarized light and subsequent heating, as described in the optical study section. As can be seen, the absolute values of the coefficients in this case are lower than those of the film that was only irradiated. This is due, again, to the aggregation favored at high temperature. However, the anisotropy of the tensor increases ($d_{31} \cong 3d_{24}$), because the starting optical anisotropy is larger.

The influence of the previous photoinduced alignment of the azo chromophores on the nonlinear susceptibility tensor is evidenced also by irradiation of the Pol-PZ-CN films with unpolarized light of 440 nm, with an incidence angle of 45° with respect to the Z-axis in Fig. 4.35 (orientation 1).

In these conditions, the azo chromophores will preferably lie along the direction of light propagation [31]. The obtained second harmonic fringes were measured with P^{ω} light, for three different orientations, by rotating the sample around its normal, and they are presented in Fig. 4.39. As shown in the figure, the signal is very asymmetric for 0° (orientation 1, Fig. 4.35) and 180°, whereas, the signal obtained at 90° (orientation 2, Fig. 4.35) is almost symmetrical. These results

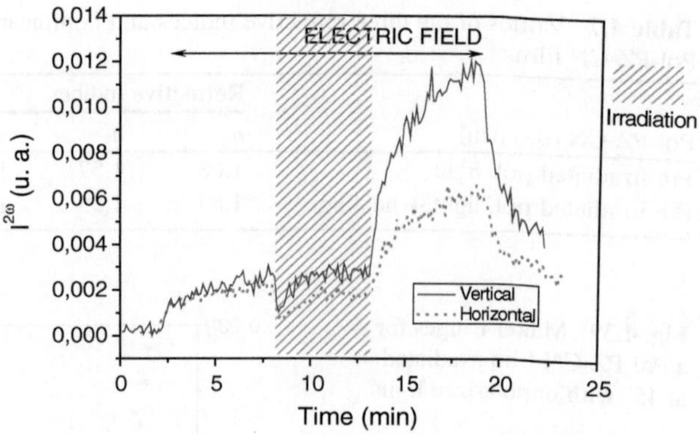

Fig. 4.40 Monitoring of the second harmonic intensity for photoassisted poling with horizontally and vertically polarized light

correspond with a polar axis generated by the effect of the electric field in the anisotropic film along a direction different from its normal.

b. In situ *irradiation (photoassisted poling)*

Finally, we examined the influence of polarized light on SHG by means of photoassisted poling. The azo chromophores are oriented in the direction of the plane perpendicular to the polarization of the incident blue light, as described by the Weigert effect. In Fig. 4.40 are shown the profiles of two photoassisted polings with 488 nm light. One of them has been obtained with vertical polarization and the other with horizontal.

As shown in this representation, when the vertically polarized blue light strikes the film, the azo chromophores are oriented perpendicular to this polarization, i.e., they are preferably arranged in the horizontal plane. Since the horizontal plane, in this case, is parallel to the P^{ω} polarization of the 1.9 μm excitation light, the measurements with P^{ω} light will give rise to a larger intensity than in the case of irradiation with horizontally polarized light, which places the chromophores in the vertical plane perpendicular to the polarization of the measurement light.

This process induces an in-plane birefringence in the films of $\Delta n \leq 0.08$. Then, the possibility was proposed that later thermal poling could also generate a biaxial response in this case. However, the coefficients found are the same as for poling performed with unpolarized light. Besides, the film was isotropic in the plane and axially oriented. This is because the difference of in-plane refractive indices achieved during the photoassisted poling process is not enough to modify the tensor symmetry. Thus, when the temperature is raised above the mesophase, the material acquires such mobility that it erases the in-plane anisotropy, aided by both the thermotropic effect and the electric field, and leading back to $C_{\infty v}$ symmetry. The cases where a biaxial NLO response is achieved, that is, those in which the in-plane anisotropy is preserved, have $\Delta n \geq 0.1$ and n_x (perpendicular to the irradiation) is slightly higher than n_z.

Table 4.8 Thermal properties of the studied copolymers

Copolymer mol % (PMMA/mon-PZ-CN)	M_n [g/mol]/PI	T_g [°C]	T_i/T_{dec}
Copol-PZ-CN_L 65/35 (70 % weight)	28000/1.5	93	132/257
Copol-PZ-CN_H 25/75 (93 % weight)	5100/1.9	104	–/237

PI: polydispersity index

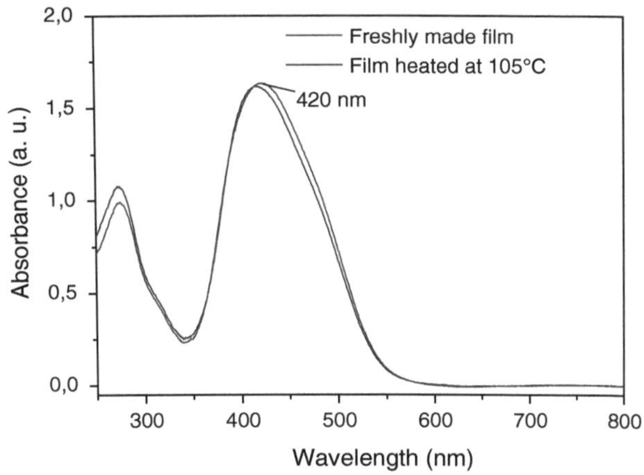

Fig. 4.41 Absorption spectra of Copol-PZ-CN_L films, freshly made and heated above its mesophase transition temperature

4.2 Copolymers

The problems found when trying to orient the Pol-PZ-CN homopolymer efficiently in a polar way derive from its remarkable tendency to aggregate, so that in addition to the irradiation method described in the previous section, the possibility of synthesizing copolymers of this material with lower azo chromophore content was considered. The copolymers that it proved possible to synthesize and their thermal properties are summarized in Table 4.8.

Both show nematic or smectic LC textures at temperatures above T_g, and the copolymer with the higher azo content decomposes before reaching the transition temperature to the isotropic state, as was the case for the homopolymer. Given the complexity of their synthesis, it was not possible to obtain a copolymer with intermediate chromophore content.

4.2.1 Optical Study

4.2.1.1 Copol-PZ-CN_L

We will begin by describing the properties of the copolymer with lower chromophore content, called Copol-PZ-CN_L. The optical absorption spectra of freshly made films are characteristic of a material with *nitrostilbene*-type azo

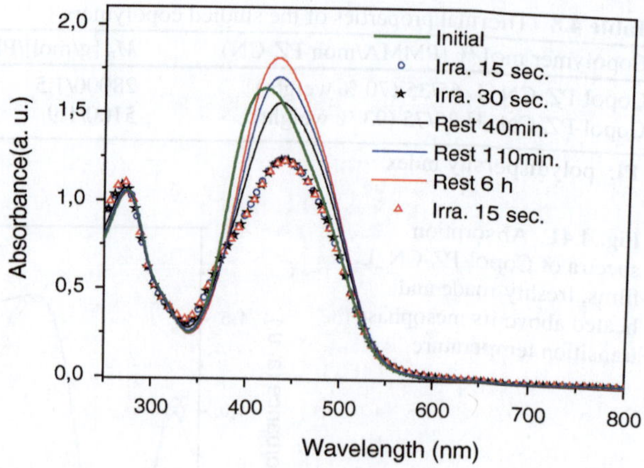

Fig. 4.42 Absorption spectrum evolution of a Copol-PZ-CN_L film under irradiation

chromophores. In Fig. 4.41 we can observe the absorption spectrum of a freshly made film and that of a film that has been heated above T_g, at 105 °C, for 60 min.

This copolymer shows a more symmetric band than the homopolymer, with its absorption maximum at 420 nm, very close to that for the solution, 430 nm. In this situation the films of this copolymer have the following refractive indices: $n_x = n_y = 1.62$, and $n_z = 1.61$, indicating that the homeotropic tendency of the homopolymer disappears when the chromophore content decreases. As shown in the figure, the heating of the film induces a small hypsochromic shift of about 4 nm, hardly noticeable, which suggests that aggregation is not as evident as in the homopolymer. Measurements of the refractive indices and absorption spectra with different polarizations and angles of incidence show that, in this state, this polymer presents no anisotropy.

The study of the effects of irradiation on the absorption spectrum, shown in Fig. 4.42, was performed starting from the situation of a film (thickness ≈0.6 μm) heated to 105 °C (initial in the figure).

Initially, the film was irradiated with unpolarized light of 440 nm wavelength for 15 s, yielding a clear decrease in the absorption band area and a small bathochromic shift of 20 nm. Two more periods of irradiation were tested, one of 30 s, and another of 2 min, which resulted in the same absorption band, indicating that the behaviour immediately reaches saturation. The successive rests led to an increase in the π-π* band area of the *trans* isomers and a slight hypsochromic displacement. Once at rest, renew irradiation results in the same band as with the first irradiation. The behaviour of the copolymer under blue light irradiation has similarities with that of the homopolymer, although there are some differences. The presence of aggregated species is evidenced in blue shift of the absorption band of the heated film. Furthermore, the increase of the band area upon irradiation indicates disaggregation, as happened for the homopolymer. However, the presence of aggregates is far less significant, since the increase of that area is not as large, and a peak centred at 390 nm is not observed. The measurement of the refractive indices for films irradiated in this way shows that the films remain practically isotropic.

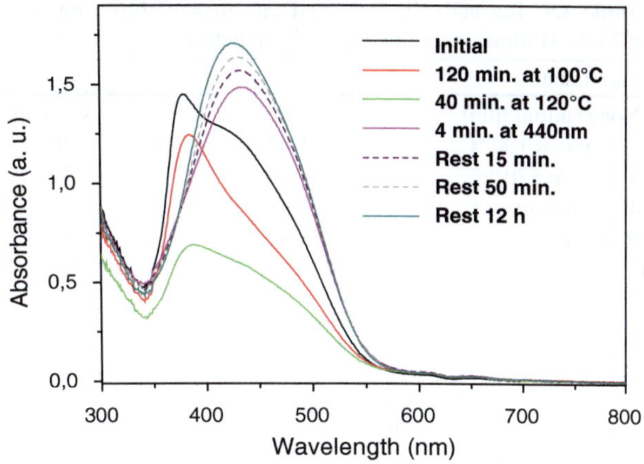

Fig. 4.43 Evolution with heating and blue light irradiation of the Copol-PZ-CN_H film of ≈ 0.7 μm thickness

Finally, treatments with polarized light and temperature were performed on the low-azo-content copolymer. A maximum photoinduced in-plane birefringence of 0.03 was achieved by irradiation. This photoinduced birefringence is not increased by further heating at $T > T_g$. Indeed, on the contrary, on surpassing T_g the tendency is again towards an isotropic situation.

It can be said, therefore, that thermal and light treatments modify the optical properties of the film to a lesser extent than in the case of the homopolymer, owing to the decrease in the amount of mesogenic photoaddressable units, which entails less aggregation.

4.2.1.2 Copol-PZ-CN_H

Copol-PZ-CN_H films were prepared and they were submitted to treatment with light and high temperature similar to that applied to the low-azo-content copolymer. The absorption spectra were recorded and are shown in Fig. 4.43. As can be seen, the behaviour is very similar to that of the Pol-PZ-CN homopolymer. The initial spectrum shows the peak that we associate with the aggregation at 390 nm. When the film is slightly heated, the band acquires more weight, and an overall decrease of the intensity is produced. On heating at higher temperature the band remains very similar to that of the homopolymer in the same conditions.

The irradiation with 440 nm light of the films in this highly aggregated state gives rise to the same spectrum and similar values of the refractive indices as the same treatment of freshly prepared film. The refractive index values associated with each treatment are gathered in Table 4.9.

As seen in Table 4.9, the homeotropic tendency of the side-chains in this case is much lower than in the homopolymer. With a relatively small reduction in the chromophore content, the birefringence of freshly prepared films obtained is significantly lower ($\Delta n = 0.02$) than that of Pol-PZ-CN ($\Delta n = 0.09$).

Table 4.9 Refractive indices and out-of-plane birefringence values (at 633 nm) of the Copol-PZ-CN_H films under heating and irradiation

Treatment	$n_x = n_y$	n_z	Δn
None (initial film)	1.63	Wavelengh(nm) 1.65	0.02
120 min at 100 °C	1.61	1.70	0.09
40 min at 120 °C	1.56	1.77	0.20
4 min 440 nm	1.65	1.71	0.06
Rest for 12 h	1.65	1.71	0.06

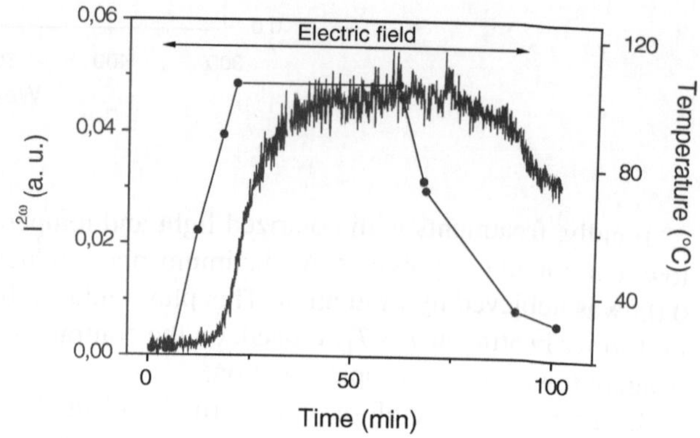

Fig. 4.44 Evolution of the second harmonic intensity generated by thermal poling for the Copol-PZ-CN_L film

By means of irradiation with linearly polarized light, using a similar process to that described for Pol-PZ-CN in Sect. 4.1.1.3, an in-plane birefringence of $n_x - n_y = 0.09$ and out-of- plane of $n_z - n_y = 0.06$ can be induced. From this state, by heating for a few seconds at 110 °C, an increase of in-plane Δn to 0.24 is achieved, which is similar to the maximum out-of-plane value obtained by the thermotropic effect for this material (Table 4.9).

4.2.2 Nonlinear Response

We will begin by describing the results of the SHG measurements after poling, obtained for Copol-PZ-CN_ L.

4.2.2.1 Copol-PZ-CN_L

Thermal poling was applied to films (≈0.6 μm thick) starting from two states: freshly prepared and pre-irradiated. The temperature was raised to 105 °C and held there for 40 min, before the film was allowed to cool to room temperature

Table 4.10 Stable values (two weeks after the orientation process) of the nonlinear coefficients, refractive indices and birefringence (at 633 nm) for the Copol-PZ-CN_L film

	d_{ij} [pm/V], recently oriented			d_{ij} [pm/V], 2 weeks		
	d_{33}	d_{31}	Δn	d_{33}	d_{31}	Δn (633 nm)
As-prepared	16	2.6	0.17	13	1.9	0.11
Pre-irradiated	17	2.6	0.18	13	1.9	0.14

in the presence of the electric field. The poling profile is illustrated in Fig. 4.44. It is observed here that the initial slope is higher than that obtained for Pol-PZ-CN, probably owing to the small amount of aggregation.

The values of the d_{33} and d_{31} coefficients measured soon after poling and two weeks thereafter, as well as the birefringence values, are summarized in Table 4.10.

From the values of Table 4.10, it follows that the previous irradiation process hardly affects the coefficients, or the anisotropy, as could be expected given the low presence of aggregates. The coefficient values are high and very stable. After a month, the measured coefficients are $d_{33} = 11$ pm/V and $d_{31} = 1.5$ pm/V. It can be checked that a decrease in the azo content with respect to the homopolymer not only does not result in a decrease of the d_{ij} values, but in fact produces an increase. The out-of-plane anisotropy decreases with the signal erasure, which in turn implies that much of it has a polar nature. Moreover, although the value of the ratio between the two coefficients is larger than three, as would be expected for a liquid crystal, it is much lower than that for the Pol-PZ-CN homopolymer. As can be deduced from the optical study, a very important difference is that the thermotropic tendency is smaller than in the homopolymer. Since the side-chains do not tend to align in an antiparallel way as clearly as in the homopolymer, aggregation, which would imply a more compact antiparallel arrangement, does not occur so extensively, and this is why the out-of-plane birefringence is so stable, while in the copolymer it drops, with the disappearance of the second harmonic signal. Therefore, the lower ratio of the two coefficients is possibly related to the lower thermotropic tendency, which in the homopolymer produces aggregation and a denser axial chromophore arrangement with a narrower angular distribution. Thus, it can be concluded that by reducing the amount of chromophore to 70 % by weight we have access to a larger number of chromophores to be oriented in a polar way, resulting in a very high NLO response and, moreover, the high coefficient ratio (d_{33}/d_{31}) characteristic of the increased order that leads to a liquid crystal is kept.

The thermal stability of the polar order induced by electric field at high temperature has been studied.

The second harmonic signal drop with temperature is shown in Fig. 4.45. As can be seen, the signal is stable up to the proximity of T_g, where it starts to decrease and after 10 min is completely erased. In these films, after the polar order "erasure", responsible for the NLO response, a certain amount of out-of-plane birefringence is still measured, this, although considerably smaller than before this heating, indicates that some of the side-chains are axially oriented. Again we find that not all the obtained axial order can be related to the NLO response, although this situation is much less pronounced than in the homopolymer.

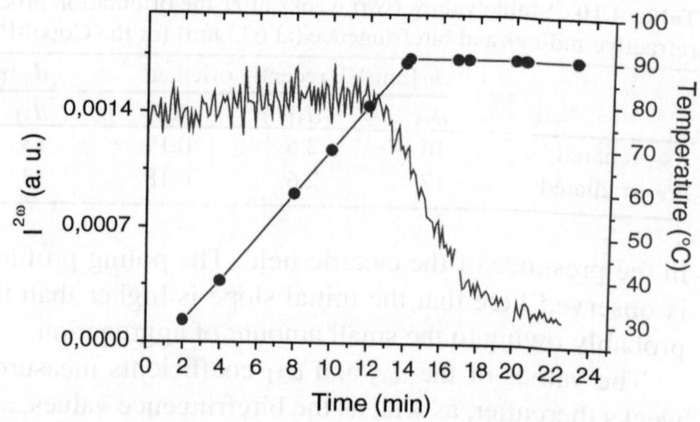

Fig. 4.45 Thermal stability of the generated second harmonic signal in a Copol-PZ-CN_L film

Photoassisted polings were performed on Copol-PZ-CN_L films, resulting in an increase of the nonlinear response at room temperature (RT) larger than for the Pol-PZ-CN homopolymer. The problems associated with the low stability of the induced order can be avoided if the photoassisted process is carried out at higher temperature.

4.2.2.2 Copol-PZ-CN_H

In general, the behaviour of Copol-PZ-CN_H is very similar to that of the homopolymer, as shown in the study of the optical properties of freshly made films and those treated with light and temperature. The poling profile this material shows is very close to that of Pol-PZ-CN, so the figure is not included here.

Figure 4.46 shows the absorption spectra for the freshly prepared film, after it has been irradiated and for two different areas of the film after thermal poling, one of which has been oriented by high temperature and in an electric field, and the other away from the field but subjected to heating.

The figure shows that the absorption band corresponding to the area affected by the electric field undergoes a hypsochromic shift compared to its previous state (irradiated). This displacement is smaller than the one obtained for the band measured in the zone which was not oriented by the field, so the aggregation appears to be lower in the presence of the electric field, and also lower than in the case of the homopolymer.

The Maker fringes were measured in order to obtain the nonlinear coefficients. The d_{ij} results obtained for thin films (0.6 μm) recently oriented by poling at 120 °C and after two weeks are shown in Table 4.11. The estimated error in these measurements is ±15 %.

As shown in Table 4.11, the nonlinear coefficients achieved by these freshly prepared copolymer films slightly improve on the ones obtained for the homopolymer, both for recently oriented samples and after two weeks. Moreover, the value obtained for pre-irradiated samples is higher and more stable. In all the above cases, the copolymer shows a d_{31} coefficient higher than that of the homopolymer,

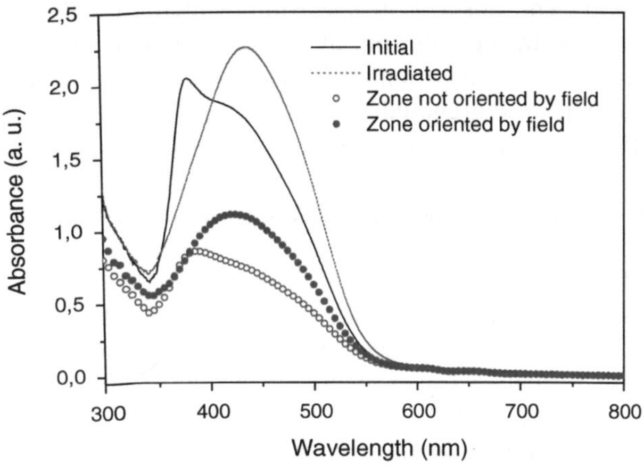

Fig. 4.46 Absorption spectra of Copol-PZ-CN_H films: freshly prepared, pre-irradiated and oriented either by thermal poling or only by high temperature

Table 4.11 Nonlinear coefficient values measured in thin Copol-PZ-CN_H films. Values for recently oriented films and after two weeks are given

	d_{ij} [pm/V] Recently oriented			d_{ij} [pm/V] + 2 weeks		
	d_{33}	d_{31}	Δn (633 nm)	d_{33}	d_{31}	Δn (633 nm)
Freshly prepared + poling	13	1.6	0.12	9	1.5	0.07
Pre-irradiated + poling	19	1.6	0.24	17	1.5	0.24

as was also the case with the one with low-azo content, so that the d_{33}/d_{31} ratios are lower, which is associated with a wider distribution of the axial order, probably due to the presence of the PMMA comonomer, which makes the packing of the side-chains more difficult. The birefringence results for Copol-PZ-CN_H show that part of the axial order achieved in the samples oriented by the usual thermal poling is due to the polar order, because it decreases when the second harmonic signal drops in time. On the other hand, in the pre-irradiated sample there is an important part of Δn that is not assignable to the polar order as it remains constant even when part of the second harmonic signal is lost. This higher axial order achieved with pre-irradiation also implies an increase in the d_{33}/d_{31} ratio with respect to the sample without irradiation. Finally, it is worth noting the large stability found in the d_{ij} values.

4.3 Conclusions

The optical absorption spectra of the homopolymer and the studied copolymers based on the mon-PZ-CN donor-π-acceptor chromophore are characteristic of *azobenzene*-type molecules, as a result of the presence of piperazine as a relatively strong acceptor. The thin films of the homopolymer, which has a smectic mesophase, are anisotropic before any treatment, with a clear homeotropic trend. The absorption spectrum, which now is shifted to higher energies compared with the solution, also shows the presence of aggregates in the starting films, preferably oriented in a direction normal to the plane containing the film. These aggregated species, whose formation and out-of-plane orientation is favoured by raising the temperature, absorb less than the isolated chromophores.

The irradiation with UV-blue light disrupts the aggregated species to some extent, as indicated by the absorption spectrum displacement to larger wavelengths of the irradiated films and the noticeable increase in intensity. Disaggregated films evolve over time towards a greater state of aggregation, showing a gradual hypsochromic shift of the absorption spectrum. The initial state of aggregation is a result of the film's formation (by evaporation) dynamics, because if we start from any state of the film (irradiated, heated or oriented by thermal poling), the re-dissolution and subsequent evaporation leads to a film that has the same initial spectrum.

It has been proved that it is possible to control, by means of light and thermal treatments, the orientation of the chromophores in the Pol-PZ-CN films, moving from a distribution that is homeotropic and isotropic in the plane, to another with values of in-plane birefringence up to $\Delta n_{\text{in-plane}} = 0.3$.

Regarding the NLO properties of the Pol-PZ-CN homopolymer, it has been possible to induce stable polar order, but the d_{33} value is significantly lower than what would be expected by considering only the $\mu\beta$ value that was measured for the monomer solution precursor. This is attributed to the substantial presence of aggregates, favoured by the interaction between chromophores leading to nonpolar entities, and which therefore do not contribute to the NLO response induced by the field. The comparison with Pol-O-CN, another azopolymer but with a weaker donor and therefore with a lower dipole moment, led to the finding that the tendency to form "stable" aggregates is smaller and the polar orientation of a larger percentage of the chromophores is possible. However, aggregation is likely to contribute to improve the temporal stability of the polar order induced in Pol-PZ-CN, as it was much better than that observed in Pol-O-CN.

As for the effect of light on the electric field orientation, it proved possible to induce a noticeable nonlinear response by photoassisted poling processes at temperatures below T_g. However, the values of the coefficients obtained are lower and less stable than those obtained by a thermal poling, because in this case the rearrangement of the polymer chains around the chromophores oriented at low temperature does not occur.

The pre-illumination of the films with UV-blue unpolarized light ostensibly improves the NLO response, as it unravels the aggregates and generates isolated chromophores, which are capable of being oriented by the electric field. If we add a short period of illumination in the first phase of the thermal poling to this pre-illumination, we obtain the optimal NLO response for the Pol-PZ-CN homopolymer.

The d_{33}/d_{31} ratios found for all cases are much higher than three. Since the electric fields used are not very intense, the high out-of-plane anisotropy of this polymer, associated with its LC nature, relates to this fact.

The orientation by thermal poling of pre-irradiated films with linearly polarized UV-blue light resulted in samples with biaxial NLO properties. The susceptibility tensor symmetry obtained has C_{2v} symmetry, which differs from that obtained by the standard methods, $C_{\infty v}$.

Random copolymers with different azo chromophore contents gave rise to less aggregated films with less homeotropic tendency. This means that the nonlinear response is closer to the results predicted by the isotropic model, particularly in the polymer with the lowest azo content, Copol-PZ-CN_L. The d_{33} coefficient values are comparable to the case of the homopolymer, with a similar thermal stability without requiring pre-irradiation. Therefore, the dilution of the chromophore content in the material is a good method to prevent aggregation and to improve significantly the NLO properties of the material.

References

1. H. Menzel et al., Small-angle X-ray scattering and ultraviolet-visible spectroscopy studies on the structure and structural changes in Langmuir–Blodgett films of polyglutamates with azobenzene moieties tethered by alkyl spacers of different length. Langmuir **10**(6), 1926 (1994)
2. J. Stumpe, T. Fischer, H. Menzel, Langmuir–Blodgett films of photochromic polyglutamates. Relation between photochemical modification and thermotropic properties. Macromolecules **29**(8), 2831 (1996)
3. T. Geue, A. Ziegler, J. Stumpe, Light-induced orientation phenomena in Langmuir–Blodgett multilayers. Macromolecules **30**(19), 5729 (1997)
4. F. Lagugné Labarthet et al., Spectroscopic and optical characterization of a series of azobenzene-containing side-chain liquid crystalline polymers. Macromolecules **33**(18), 6815 (2000)
5. S. Freiberg et al., Investigation of thermochromism in a series of side-chain, liquid-crystalline, azobenzene-containing polymers. Can. J. Chem. **82**(1), 1 (2004)
6. V. Rodriguez et al., Quantitative determination of the polar order induced under high electric field in amorphous PDR1M azobenzene polymer films. J. Phys. Chem. B **107**, 9736 (2003)
7. A.M. Kelley, A multimode vibronic treatment of absorption, resonance Raman, and hyper-Rayleigh scattering of excitonically coupled molecular dimers. J. Chem. Phys. **119**(6), 3320 (2003)
8. H. Rau, in *Photochemistry and Photophysics*, Vol. 2, ed. by J. F. Rabek (CRC Press, Boca Ratón, FL, 1990), Chap. 4
9. A. Natansohn, P. Rochon, Photoinduced motions in azo-containing polymers. Chem. Rev. **102**(11), 4139 (2002)
10. F. J. Rodríguez, *Propiedades ópticas fotoinducidas en polímeros con uidades de azobenceno* (2005) Ph.D. Dissertation, Universidad de Zargoza
11. C. Sánchez et al., Biphotonic holographic recording in a liquid crystalline cyanoazobenzene side-chain polymethacrylate. Polarization, intensity, and relief gratings. J. Appl. Phys. **89**(10), 5299 (2001)
12. H.L. Hampsch et al., Second harmonic generation in corona poled, doped polymer films as a function of corona processing. J Appl Phys **67**(2), 1037 (1990)
13. H. Wang, R.C. Jarnagin, E.T. Samulski, Electric field poling effects on the molecular reorientational dynamics of side-chain nonlinear optical polymers. Macromolecules **27**(17), 4705 (1994)
14. T. Weyrauch in *Relaxation Phenomena. Liquid Crystals, Magnetic Systems, Polymers, High-Tc Superconductors, Metallic Glasses*, ed. by W. Haase, S. Wróbel (Springer, Berlin, Heidelberg, 2003)

15. N. Tirelli et al., Structure–activity relationship of new organic NLO materials based on push-pull azodyes. 1. Synthesis and molecular properties of the dyes. J. für praktische Chemie **340**(2), 122 (1998)
16. M. Makowska-Janusik et al., Molecular dynamics simulations of electric field poled nonlinear optical chromophores incorporated in a polymer matrix. J. Phys. Chem. B **108**(2), 588 (2004)
17. Z. Sekkat, M. Dumont, Photoassisted poling of azo dye doped polymeric films at room temperature. Appl. Phys. B **54**(5), 486 (1992)
18. Z. Sekkat et al., Room-temperature photoinduced poling and thermal poling of a rigid main-chain polymer with polar azo dyes in the side chain. Chem. Mater. **7**(1), 142 (1995)
19. Z. Sekkat et al., Correlation between polymer architecture and sub- glass-transition-temperature light-induced molecular movement in azo-polyimide polymers: influence on linear and second- and third-order nonlinear optical processes. J. Opt. Soc. Am. B **15**(1), 401 (1998)
20. D. J. Williams, in *Nonlinear Optical Properties of Organic Molecules and Crystals*, Vol. 1, ed. by D. S. Chemla, J. Zyss (Academic, New York, 1987)
21. M. Eich et al., Corona poling and real-time second-harmonic generation study of a novel covalently functionalized amorphous nonlinear optical polymer. J. Appl. Phys. **66**(6), 2559 (1989)
22. V. Rodriguez, F. Lagugné-Labarthet, C. Sourisseau, Orientation distribution functions based upon both < P1 >, < P3 > order parameters and upon the four < P1 > up to < P4 > values: application to an electrically poled nonlinear optical azo polymer film. Appl. Spectrosc. **59**(3), 322 (2005)
23. J.C. Dubois et al., Behaviour and properties of side chain thermotropic liquid crystal polymers. Acta. Polym. **48**(3), 47 (1997)
24. M. Amano et al., Second order nonlinear optical properties of polymers containing mesogenic side chains. Mol. Cryst. Liq. Cryst. Incorporating Nonlinear Optics **182**, 81 (1990)
25. D. Gonin et al., Side chain liquid crystalline polymers: electric field effects and nonlinear properties. Macromol. Symp. **96**(1), 185 (1995)
26. P.M. Blanchard, G.R. Mitchell, A comparison of photoinduced poling and thermal poling of azo-dye-doped polymer films for second order nonlinear optical applications. Appl. Phys. Lett. **63**(15), 2038 (1993)
27. X. Meng et al., Azo polymers for reversible optical storage. 10. Cooperative motion of polar side groups in amorphous polymers. Macromolecules **29**(3), 946 (1996)
28. T. Fischer et al., Interdependence of photoorientation and thermotropic self-organization in photochromic liquid crystalline polymers, Molecular Crystals and Liquid Crystals Science and Technology. Section A. Molecular Crystals and Liquid Crystals 298 (1997):213
29. J.C. Kim et al., Control of three-dimensional refractive indices by both drawing and poling of functionalized phenoxy side-chain polymers. Macromolecules **29**(22), 7177 (1996)
30. X.T. Tao et al., Phase-matched second-harmonic generation in poled polymers by the use of birefringence. J. Opt. Soc. Am. B **12**(9), 1581 (1995)
31. K. Ichimura, M. Han, S. Morino, Photochemistry determined by light propagation. part 1: three-dimensional photomanipulation of self-organized azobenzenes in liquid-crystalline polymers. Chem. Lett. **28**(1), 85 (1999)

Chapter 5
Films of Doped Low Polar Azopolymers

In this part of the thesis, a study of the nonlinear optical (NLO) properties of films composed of highly efficient chromophores dispersed in a photoaddressable liquid-crystalline polymer (LCP) matrix is presented.

A common method of creating polymeric materials with NLO response is to disperse NLO chromophores in a polymeric matrix. While the chromophores provide the desired nonlinear properties, the matrix determines other properties, such as processability and both mechanical and thermal behaviour. The orientation of these mixtures by means of "thermal poling" is a process often used for materials lacking a centre of symmetry and therefore suitable for second harmonic generation (SHG) [1, 2]. If photoaddressable chromophores are used in such systems, the high temperature at which they undergo orientation can be changed by irradiation with a suitable wavelength, so-called "photoassisted poling", which, as described in Chap. 4 of this thesis, involves the photoinduction of *trans-cis-trans* isomerization simultaneously to the application of an electric field. The difference now is that in Chap. 4 the chromophore, forming the side chain, had the azobenzene group, and it was therefore liable to isomerize. Most work on guest–host mixtures has been carried out using photoaddressable chromophores, as in the case of the system composed of PMMA and DR1 dispersed or attached to the main chain [3, 4].

However, in the present case, the photoaddressability properties will be delegated to the polymer matrix instead of to the nonlinear chromophore. Thus, we will study mixtures of not photoisomerizable chromophores with matrices, as an alternative to single-component systems. One of the advantages of these systems is the possibility of using highly efficient nonlinear chromophores and exploiting the benefits of photoassisted processes for polar orientation while avoiding the inconvenience of simultaneously optimizing both properties in a single system, which usually results in significant difficulties in the synthesis. The first and almost only approach to these systems was described by Ishow et al. [5]. In their work, they used *N*-(4-nitrophenyl)-L-prolinol (NPP) as the chromophore, and two azocopolymers with such a low nonlinear response that it could be neglected. They concluded that the excitation of the polymer matrix by means of light irradiation

R. Alicante, *Photoinduced Modifications of the Nonlinear Optical Response in Liquid Crystalline Azopolymers*, Springer Theses, DOI: 10.1007/978-3-642-31756-9_5,

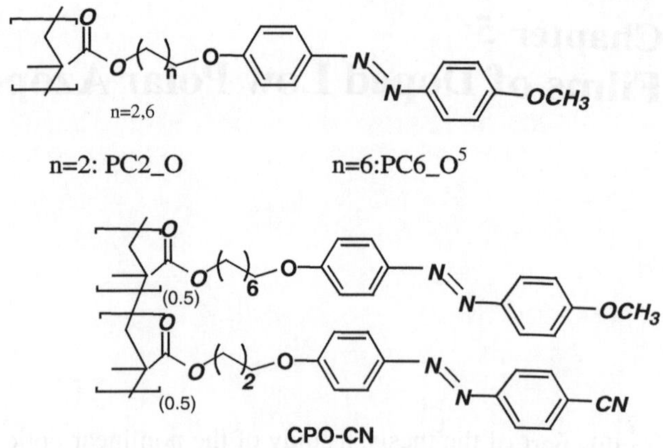

Fig. 5.1 Chemical structures of the studied azopolymers [9]

facilitates the polar orientation of the dispersed chromophores. More recently, Dalton et al. [6–8]., in their study on the effects of laser-assisted polar orientation in binary systems, have suggested that the modification of the matrix order by photoassisted poling significantly affects the final order of the dispersed chromophore, finding also a noticeable increase in the electrooptic coefficient value (r_{33}) with respect to the purely thermal process. Therefore, the main objective of the work reported in this chapter was to study the influence of the interactions between the chromophores and the matrix in the generated polar order in azopolymeric films with dispersed chromophores. For that, we will analyse the second harmonic intensity generated by these films after have undergone both thermal and photoassisted processes of polar orientation.

We begin with a brief description of the optical properties of the components of the matrix/chromophore systems to be studied. In particular, the arrangements that the polymers used as a matrix acquire when they are subjected to optical and/or thermal treatments will be analysed. For this purpose, UV-vis absorption spectroscopy and refractive index measurements will be used. Then, the nonlinear response of the systems oriented under different conditions will be measured, and the characteristics of the mentioned response will be analysed as a function of the induced order in the matrix.

5.1 Description of the Guest–Host (Chromophore–Polymeric Matrix) Systems

The polymers used as photoaddressable matrices in this study are shown in Fig. 5.1. Initially, two side chain liquid-crystalline (LC) azopolymers with weak donor and acceptor groups were chosen. On the other hand, a random LC copolymer was used, which bears azochromophores with a stronger acceptor group in half of its side chains. This copolymer is included to study the influence on the

Table 5.1 Molecular weights and transition temperatures of the polymers

Polymer	M_n [g/mol]/PI	Thermal transitions	$T_{dec.}$
PC6_O	25400/2.1	g 64 S 89 N 126 I	359
PC2_O	14000/2.1	g 111 N 157 I	272
CPO-CN	13675	g 74 N 142 I	347
PMMA	83000	g ≈ 100	–

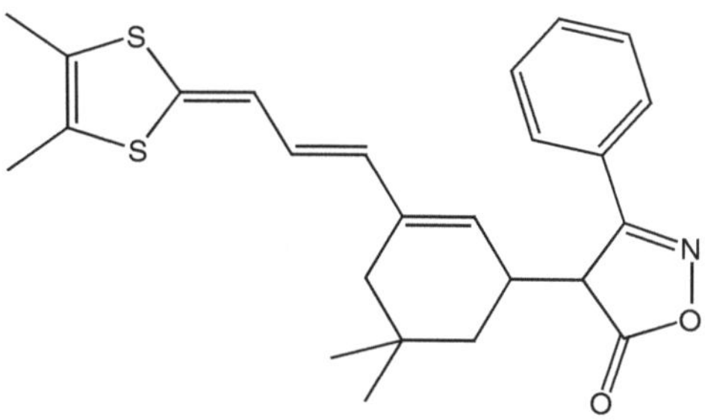

Fig. 5.2 Chemical structure of the studied chromophore, Me-3-FX

intermolecular interactions of a higher dipole moment in the side chain of the matrix. At the same time, for the sake of comparison, films of the chromophore mixed with amorphous PMMA as an inert matrix (not photoaddressable) were prepared, with a molecular weight of $M_n = 80000$ g/mol.

The chemical structures of the two azo homopolymers have a methylene spacer with different length between the azobenzene and the main chain. The copolymer is composed of 50 % of the monomer of the PC6_O polymer and 50 % of the monomer of PC2_O but with the methoxy group replaced by a cyano group.

These structural differences translate into different thermal properties, which are summarized in Table 5.1. As shown in the table, the azopolymers show typical transitions of a LC nature. PC2_O and the CPO-CN copolymer showed nematic LC textures above the glass transition temperature T_g, while for PC6_O a smectic phase is observed above T_g and a nematic one above about 90 °C.

Regarding the incorporated nonlinear chromophores, in the preliminary studies, films with different molecules selected from among those studied in Chap. 3 were prepared. Some of the mixtures did not give rise to films with sufficient optical quality, as happened with the mixtures made with the chromophore **28**, which has the diphenylpyran group as a donor and tricyanofuran as the acceptor. In other cases, such as films doped with **26bii** and **26bi**, they did not properly withstand light treatment, and a photodegradation of the chromophore was clearly observed. This was true also for **35b** and for another similar compound with a dithiol group as the donor. Some mixtures with the chromophore **19c** were also prepared; however, the response to thermal poling showed very poor reproducibility, so its use was discontinued. Finally, we selected **14a** [10] as the main chromophore of the study, which we will refer to as Me-3-FX in this chapter. Compound **14a** has the structure shown in Fig. 5.2.

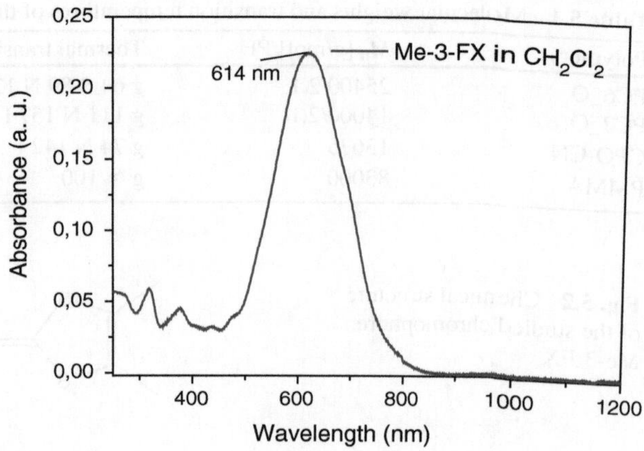

Fig. 5.3 Absorption spectrum of the Me-3-FX chromophore in CH_2Cl_2 solution

It is a merocyanine type of chromophore, whose acceptor group is an oxazolone and the donor group a dithiol, both with proaromatic character. The spacer is an isophorone group, which improves the thermal and photooxidative stability with respect to the analogous linear polyene compound (see Chap. 3). This compound shows an acceptable thermal stability at the temperatures employed in the orientation process.

5.2 Optical Study

The Me-3-FX chromophore to be used does not show any absorption at the 1907 nm excitation wavelength and none at the 954 nm second harmonic either. The absorption at 406 nm (wavelength to be used for the photoassisted poling) is very small. As discussed in Chap. 3, it shows positive solvatochromism, that is, the absorption maximum is shifted to the red when a more polar solvent is used. The absorption spectrum in dichloromethane solution is shown in Fig. 5.3.

5.2.1 Azopolymer Films

Films of the three azopolymers were prepared by "casting". The photoadressability of the matrices was checked by subjecting freshly made films of PC2_O, CPO-CN and PC6_O to irradiation with light of 406 nm wavelength. The absorption spectra revealed the existence of a *trans-cis-trans* isomerization. The evolution is very similar for all three polymers, except for the fact that for PC2_O and PC6_O homopolymers the *cis* isomers have a longer half-life than in the case of the copolymer with the cyano group, a stronger acceptor. In Fig. 5.4a the evolution of the absorption spectrum with irradiation for CPO-CN copolymer is shown. The optical characterization and behaviour under irradiation of PC6_O and PC2_O

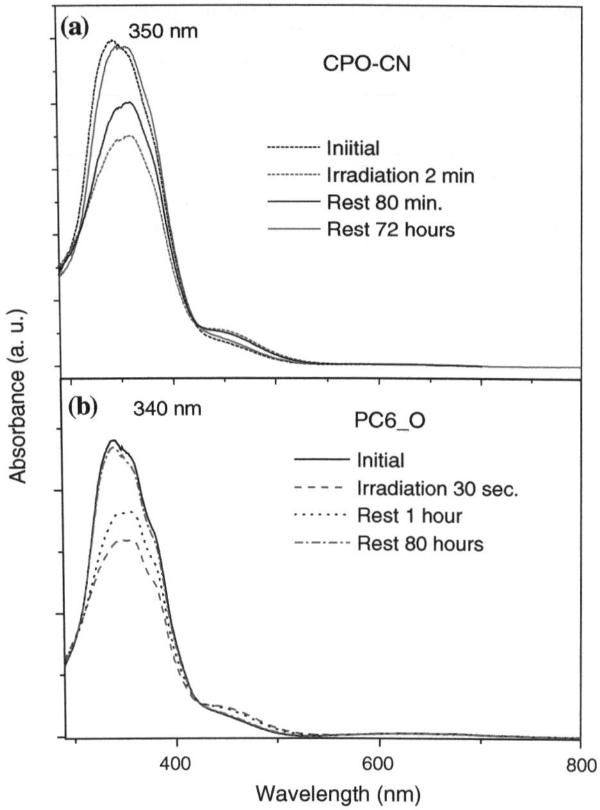

Fig. 5.4 Absorption spectra for the evolution of films of CPO-CN (**a**) and PC6_O (**b**) under irradiation with light

materials have been previously described by absorption spectroscopy and the results reported in several publications (Fig. 5.1) [11–13]. Figure 5.4b shows that the behaviour we measured for PC6_O film, very similar to that of PC2_O, agrees with the reported data [11].

Irradiation of CPO-CN with 406 nm laser light induces a *trans-cis-trans* isomerization, revealing a band at 460 nm, while the band centred at about 350 nm decreases in intensity. At rest the initial band recovers, and the one assigned to the *cis* species disappears. The small difference in the absorption maximum for freshly made CPO-CN film and that measured at rest after irradiation is related to the existence, not very significant, of aggregates in the initial state.

The irradiation of films of these materials with polarized light leads to photoinduced anisotropy, which was characterized by measuring the refractive indices.

The values of the refractive indices for freshly made and irradiated films are shown in Tables 5.2 and 5.3.

As can be seen, the refractive indices of the films were measured at two different wavelengths to obtain information about the refractive index dependence on the wavelength, since it is necessary to know them at 954 nm and 1,907 nm

Table 5.2 Refractive indices, birefringence and average index for the three azopolymer films

	As-prepared films							
	633 nm				1306 nm			
Polymer	$n_x = n_y$	n_z	$\Delta n_{out\text{-}of\text{-}plane}$	n_{av}	$n_x = n_y$	n_z	$\Delta n_{out\text{-}of\text{-}plane}$	n_{av}
PC6_0	1.63	1.63	0	1.63	1.59	1.58	−0.01	1.59
PC2_0	1.64	1.63	−0.01	1.64	1.59	1.58	−0.01	1.59
CPO-CN	1.65	1.63	−0.02	1.64	1.61	1.59	−0.02	1.60

Table 5.3 Refractive indices, birefringence, and average index for the irradiated films

	Irradiated films				
	633 nm				
Polymer	n_x	n_y	n_z	$\Delta n_{in\text{-}plane}$	n_{av}
PC6_0	1.64	1.62	1.64	0.02	1.63
PC2_0	1.66	1.63	1.65	0.03	1.65
CPO-CN	1.65	1.62	1.65	0.03	1.64

Table 5.4 Refractive indices, birefringence and average index for the three azopolymer films

	Thermally treated films							
	633 nm				1306 nm			
Polymer	$n_x = n_y$	n_z	$\Delta n_{out\text{-}of\text{-}plane}$	n_{av}	$n_x = n_y$	n_z	$\Delta n_{out\text{-}of\text{-}plane}$	n_{av}
PC6_0	1.54	1.82	0.28	1.64	1.51	1.74	0.23	1.59
PC2_0	1.68	1.56	−0.12	1.64	1.61	1.53	−0.08	1.58
CPO-CN	1.55	1.82	0.27	1.65	1.54	1.73	0.19	1.61

in order to obtain the nonlinear coefficients. For this purpose, fitting of the data in the table and measurements of the indices at 780 nm was performed using the Sellmeier expression as shown in Fig. 4.7.

In order to analyse the orientation that these films attain as a result of the thermotropic effect, they were subjected to an "annealing" process, i.e. they were heated to a temperature in the nematic mesophase (95 °C for PC6_O and CPO-CN, 120 °C for PC2_O) for a few minutes. The refractive indices at 633 nm and 1,306 nm are shown in Table 5.4. As can be seen in the tables, the behaviour is different in terms of the arrangement acquired by the polymer side chains. PC6_O and CPO-CN are oriented homeotropically or perpendicularly to the film surface, while in PC2_O the side chains are oriented parallel to the film plane, reaching a lower absolute value of the birefringence. Moreover, the films remain isotropic in the plane ($n_x = n_y$) in all cases except when they are irradiated with linearly polarized light.

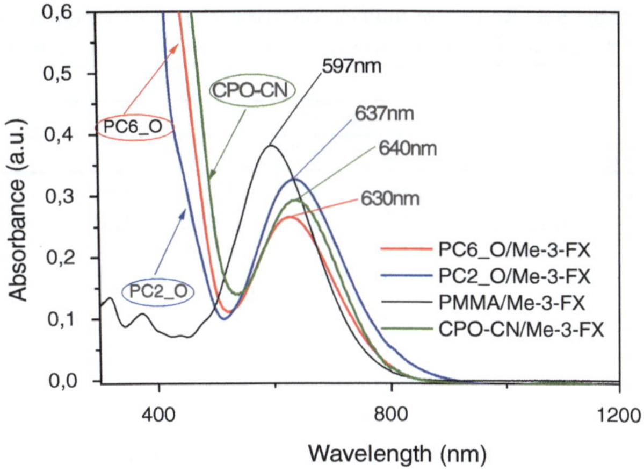

Fig. 5.5 Absorption spectra of the azopolymer and PMMA/chromophore films

5.2.2 Azopolymer/Chromophore Films

Once we had characterized the polymer films, films doped with the chromophore were prepared. The chromophore content was approximately 2.5 % by weight for the four chromophore/polymer mixtures: PMMA/Me-3-FX, PC2_O/Me-3-FX, PC6_O/Me-3-FX and CPO-CN/Me-3-FX. The area of the spectrum that corresponds to the chromophore absorption in these mixtures is presented in Fig. 5.5. As can be seen in the spectra, the chromophore dispersed in PMMA shows a small hypsochromic shift in solution, whereas when it is dispersed in the azopolymers the absorption maximum is slightly shifted to lower energies. This behaviour is due to the greater dipole moment that the chromophore shows in the charge-transfer (CT) state with respect to the fundamental one, so a polar environment such as the azopolymer matrix stabilizes the excited state by decreasing the energy considerably, thus there is positive solvatochromism. In a nonpolar environment, such as PMMA, the ground state is favoured and a slight hypsochromic shift takes place.

The three mixtures were subjected to analogous treatments to those used in the optical study of pure azopolymers. The values of the refractive indices for the films of the three mixtures are shown in Tables 5.5 and 5.6.

The indices were measured only at 1,306 nm, since at 633 nm the high absorption of the films prevented us from obtaining results. Also, the mixture with PMMA (n_{av} (PMMA) = 1.49) showed a refractive index below that of the substrate, which could not be determined by this technique.

As can be seen, the average refractive index is slightly larger than that of the pure polymers. Furthermore, it was found that by increasing the chromophore content in the films the average index value also becomes higher. However, the thermotropic trend characteristic of each matrix remains.

Table 5.5 Refractive indices, birefringence and average refractive index for the three azopolymer/chromophore films, freshly made and subjected to thermal treatment

	1306 nm							
	As-prepared				Thermal treatment			
Mixture	$n_x = n_y$	n_z	Δn	n_{av}	$n_x = n_y$	n_z	Δn	n_{av}
PC6_0/Me-3-FX	1.60	1.60	0	1.60	1.53	1.76	0.23	1.61
PC2_0/Me-3-FX	1.62	1.60	−0.02	1.61	1.65	1.56	−0.09	1.62
CPO-CN/Me-3-FX	1.63	1.61	−0.02	1.62	1.57	1.69	0.12	1.61

Table 5.6 Refractive indices, birefringence and average index for the three azopolymer/chromophore films, under vertical polarized irradiation at 406 nm (*X* direction)

	1306 nm				
	Irradiated				
Polymer	n_x	n_y	n_z	$\Delta n_{in\text{-}plane}$	n_{av}
PC6_0/Me-3-FX	1.58	1.60	1.59	0.02	1.59
PC2_0/Me-3-FX	1.59	1.63	1.62	0.04	1.61
CPO-CN/Me-3-FX	1.58	1.63	1.62	0.05	1.61

Once the orientational tendency of the matrix is known, and having found that the mixtures acquire the same kind of order as the matrix under different treatments, we studied the influence of these factors on the nonlinear response generated by the highly efficient chromophore.

5.3 Nonlinear Optical Response

5.3.1 Thermal Poling

Below, we will analyse the NLO response by means of SHG measurements in chromophore-doped films previously oriented by thermal poling.

Prior to the study of films, the NLO properties of the Me-3-FX molecule in dichloromethane solution were measured at 1.9 μm, giving the result $\mu\beta = 2750 \times 10^{-48}$ esu ($\mu\beta\,(0) = 1475 \times 10^{-48}$ esu), as stated in Chap. 3, where the molecule is called **14a**. This value is very high, eight times that obtained for the chromophore mon-PZ-CN, which was the active unit of the films presented in Chap. 4. This high nonlinearity allows the study of mixtures of polymers with very low chromophore content, to avoid miscibility problems, aggregation, etc., while maintaining a high intensity in the NLO response.

The first step was to characterize the response of the chromophore dispersed in an inert polymer, orienting PMMA/Me-3-FX films by thermal poling at 95–100 °C. A thermal profile of the poling process was conducted to obtain the optimal temperature conditions, where the chromophore acquires a maximum mobility while preventing possible damage in the sample. This process, as discussed in the previous chapter, consists in monitoring the intensity of the second harmonic generated by slowly increasing the temperature of the film during corona discharge. Once the film is oriented, this process must be reproducible, i.e. the film is then heated above the isotropic phase, thus completely erasing the SHG signal, and when the same process is repeated the same results should be obtained. Once the thermal poling was completed, Maker fringes for two polarizations of the incident light were measured. The maximum value for the NLO coefficients we found 2 h after poling was $d_{33} \approx 6$ pm/V and $d_{31} \approx 2$ pm/V. The stability of this signal is very low, since the chromophore is just dispersed in the matrix, that is, it is not covalently bound to the polymer chain, and once the orienting electric field is removed it becomes free to return to its thermodynamic equilibrium position. Thus, after only 30 days the coefficients are reduced to 20 % of the initial value. Moreover, the ratio d_{33}/d_{31} is three, the characteristic value for an amorphous material in low field conditions. In this way, we ensured that the conditions we would use with the azopolymers were suitable.

An estimate of the maximum achievable coefficients from the molecular hyperpolarizability value results in the coefficients $d_{33} \approx 10$ pm/V and $d_{31} \approx 3$ pm/V. This calculation was done according to the isotropic model through the following equation, which appears in Chap. 1 of this thesis and in Appendix C:

$$d_{33} = \frac{N f_{\omega}^{2} f_{2\omega} \mu E \beta_z}{10kT}, \quad \text{where } < \cos^3\theta > \approx \frac{\mu E}{5kT}$$

$E \cong 10^8$ V/m
$N \cong 2.5 \times 10^{26}$ molec m^{-3}
$T \cong 400$ K

$\mu\beta_z = 3.84 \times 10^{-66}$ C m^5 V^{-1}
$f_{\omega}{}^2 f_{2\omega} \cong 4.6$
$K = 1.38 \times 10^{-23}$ C V K^{-1}

Through this calculation estimates of some values of the electric field were made, which were more or less accurate approximations. Besides, a perfectly homogeneous dispersion of the chromophore in the PMMA was assumed for the calculation of N. The local field factor due to PMMA was calculated assuming a dielectric constant ε equal to that obtained by H. W. Guan and C. H. Wang [14] for a mixture of 2 % of the 2-methyl-4-nitroaniline chromophore in PMMA at 95 °C. Considering the approximations made and the accuracy of the measurements, we find that although the amount of chromophore added to the system is very small, the NLO response is quite large, when compared, for example, with that of Pol-O-CN (Chap. 4), which includes 100 % of the chromophore attached to the main chain.

If we analyse the NLO response of the films of PC6_O (95 °C) and PC2_O subjected to an analogous thermal poling, it is very low due to the low hyperpolarizability of the azochromophore, almost symmetrical. The d_{33} values are at least an order of magnitude below that of the chromophore dispersed in PMMA. In this way, we ensured that the nonlinear response of the mixture we were going to measure was due essentially to the Me-3-FX chromophore dispersed in the film. It is also important to mention that in these two azopolymers there is no out-of-plane birefringence in the area affected by the electric field. Since the temperature and exposure time during thermal poling is similar to the "annealing" process that led to remarkable birefringence values (Table 5.4), it seems that the application of the corona discharge somehow hinders the thermotropic tendency of the mesogens [15].

It should be noted that the case of the CPO-CN copolymer oriented at 95 °C is slightly different, owing to the larger acceptor character of the CN group, which results in a greater SHG intensity. The orientation by electric field, in this case, gave rise to some out-of-plane orientation ($\Delta n = 0.03$) and to d_{33}/d_{31} ratios somewhat higher than three ($d_{33}/d_{31} \approx 4$). The anisotropy value does not reach that obtained by "annealing" at the same temperature and for similar heating times, which suggests that, also in this case, the electric field perturbs the thermotropic trend. On the other hand, if the poling temperature is raised above 130 °C, an out-of-plane anisotropy of 0.31 is reached and the following values for the coefficients are found: $d_{33} = 6$ pm/V and $d_{31} = 0.4$ pm/V. We note here that the d_{33}/d_{31} ratio is 15, much higher than that for amorphous polymers [16], which is typical of systems with strong anisotropy.

In examining the behaviour of the films of mixtures under "soft" thermal poling conditions, we note that PC2_O/Me-3-FX and PC6_O/Me-3-FX showed no out-of-plane birefringence, and the ratios obtained were close to three, so that in both cases the matrix would behave almost like an amorphous polymer. When the temperature of the poling process was increased, a small birefringence (Δn of 0.04–0.05) was measured and the d_{33}/d_{31} ratio reached six. As explained in Chap. 4 of this thesis, these ratios are characteristic of liquid crystals and of a larger axial orientation. Since dispersed chromophores are responsible for the SHG signal, it can be sensed that it acquires a similar orientation to that of the azopolymer. However, since the azopolymers in these mixtures do not acquire a clear homeotropy, further experiments were proposed to reveal the existence of this phenomenon.

Orientation of the CPO-CN/Me-3-FX mixture under "soft" conditions (95 °C) results in very low birefringence and d_{33}/d_{31} ratios only slightly above three. However, when the poling temperature is increased to above 130 °C, the birefringence is close to 0.3 and the coefficients are $d_{33} = 16$ pm/V and $d_{31} = 1.2$ pm/V. Both values are nearly three times those of the undoped copolymer mentioned above ($d_{33} = 6$ pm/V and $d_{31} = 0.4$ pm/V) and therefore most NLO response is attributed to the chromophores, which have been oriented in the order established by the matrix, thus showing a coupling in the movement.

In order to extend our understanding of the nature of this coupling of the motion of the two components of the system, a study of the influence of the magnitude of the azopolymer side chain's dipole moment was proposed. As we have seen, during the thermal poling of the samples with homeotropic tendency, PC6_O/Me-3-FX, this trend is hampered by the application of the electric field. However, orientation by "annealing" can be carried out prior to the induction of polar order in the films, so that before the thermal poling process we can have the matrix already oriented homeotropically ($\Delta n = 0.23$).

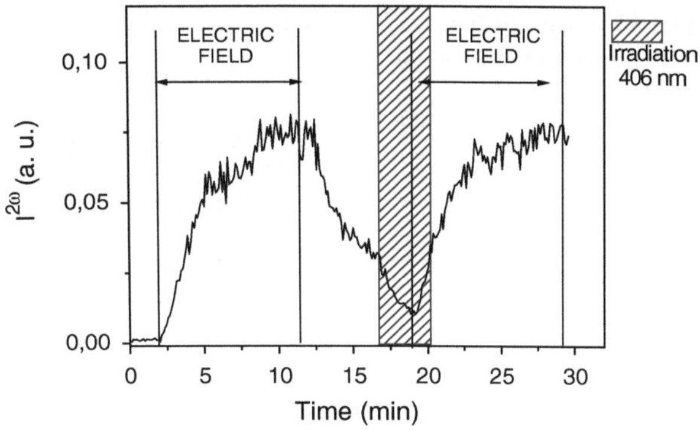

Fig. 5.6 Photoassisted poling at 30 °C of the PMMA/Me-3-FX mixture

This method results in the ratio $d_{33}/d_{31} = 6$ for PC6_O/Me-3-FX, similar to that obtained in samples oriented at high temperature. Although in both cases the matrix takes on a very different degree of order, it seems that the movement of the chromophore is similar, and that there is a maximum content of chromophore that the matrix is able to orient. If we submit the mixture with larger dipole moment (CPO-CN/Me-3-FX) to the same process, higher ratios are obtained, of about 10, which points to an increased coupling of the motion of the chromophore with that of the side chain of the copolymer, compared to PC6_O. Therefore, it can be concluded that the interactions between the side chains of the matrix and the chromophore play an important role in the final NLO properties of the mixture.

5.3.2 *Photoassisted Poling*

In this section, we will study the behaviour of the mixtures when they are oriented by means of light and electric field at low temperature. Both films of the polymers used as the matrix and mixtures of these with the chromophore were subjected to photoassisted poling processes at 30 °C. The typical process consisted of three steps: first, the electric field was applied for 10 min, then the field was removed to let the signal relax and, finally, the sample was irradiated with light and the electric field was turned on again.

The orientation of the chromophores occurs, in principle, because they are in a matrix that generates free volume when it is isomerized by irradiation, and allows the movement of the chromophores by the electric field. So, we should begin by ruling out possible effects of light (orientation or degradation of the chromophores) in the NLO response. Thus, the PMMA/Me-3-FX mixture was subjected to photoassisted poling with linearly polarized light of 406 nm wavelength. Figure 5.6 shows the in situ monitoring of the second harmonic signal during the three-step orientation process described.

As can be seen in the figure, the chromophore achieves the same intensity of generated second harmonic signal by using only the electric field at 50 °C as when

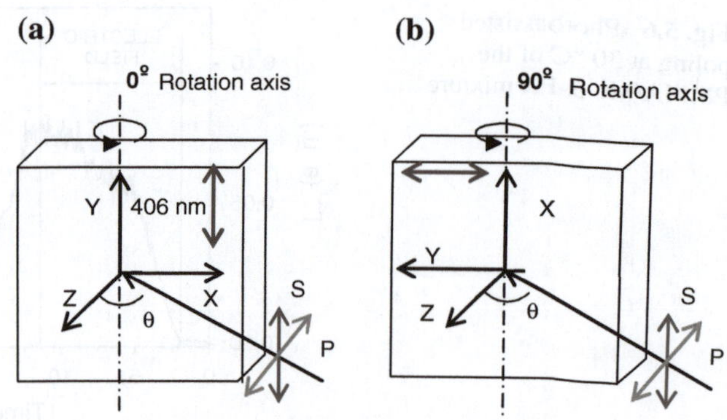

Fig. 5.7 (Same as Fig. 4.35.) Measurement configuration of Maker fringes. **a** Orientation 1, **b** Orientation 2

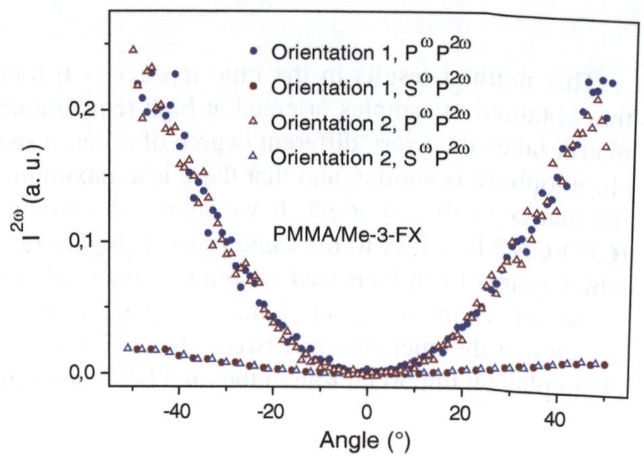

Fig. 5.8 Maker fringes for a PMMA/Me-3-FX film oriented by photoassisted poling at 50 °C. The fringes are obtained for two measurement configurations and two polarizations of the incident light

the field and illumination are applied simultaneously, so any possible effect light may induce in the chromophores does not seem to have any consequences for the SHG signal. In any case, in order to confirm this result, we performed measurements of the Maker fringes generated in this way for two orientations of the PMMA/Me-3-FX film and two polarizations of the incident light. Thus, we can see that the chromophore is not significantly altered, since it does not result in a decrease of the signal in the vertical direction (irradiation direction). The scheme of these measurements is the same as used in Chap. 4 for the Pol-PZ-CN material, and it is recalled as Fig. 5.7, which is a reproduction of Fig. 4.35.

The fringes obtained in this way present the same intensity in both orientations and are shown in Fig. 5.8.

If we now apply the same process to the three azopolymers, there is, in contrast, a clear orienting effect of the light of 406 nm wavelength. Then, Fig. 5.9 shows the polar orientation profiles obtained for PC6_O and CPO-CN in the photoassisted poling processes.

As can be observed, the behaviour under laser illumination with vertically polarized light of 406 nm wavelength is very similar in the two cases. As was noted in the case of Pol-PZ-CN (Chap. 4), light exerts a beneficial effect due to the presence of the *cis* species, which gives mobility to the material at

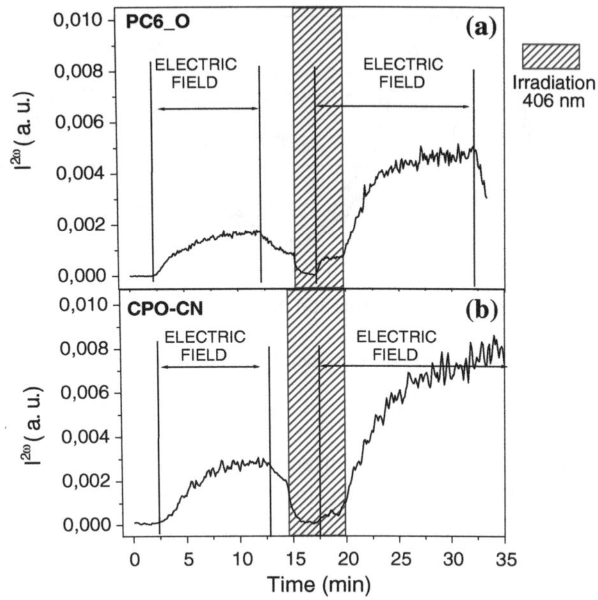

Fig. 5.9 Photoassisted poling (three-step process) for the azopolymer films PC6_O (**a**) and CPO-CN (**b**)

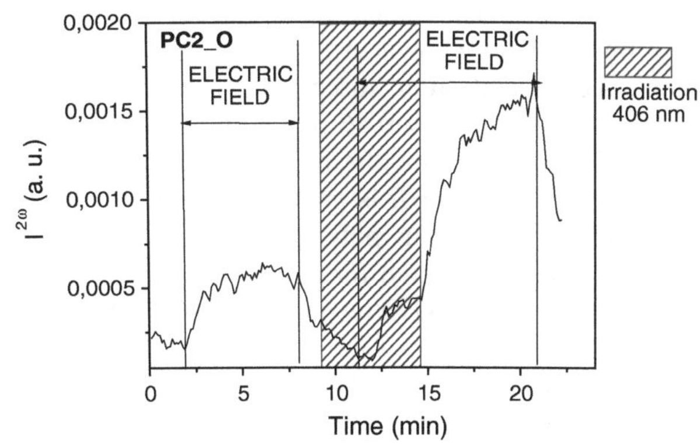

Fig. 5.10 Photoassisted poling (three-step process) for the PC2_O azopolymer films

temperatures below its T_g, and in both cases the second harmonic signal is nearly tripled with respect to the orientation induced only by the electric field. In Fig. 5.9 it is shown that the intensity is larger for the copolymer (which has a higher T_g) than for PC6_O, owing to the presence of the CN group, a strong electron acceptor group.

The PC2_O azopolymer shows the same NLO response pattern (Fig. 5.10). However, globally, results in a lower response at 30 °C than the one for PC6_O. This may be because it is farther from its T_g and the material is more rigid, so that the side chains have less mobility and are thus less easily oriented by the electric field.

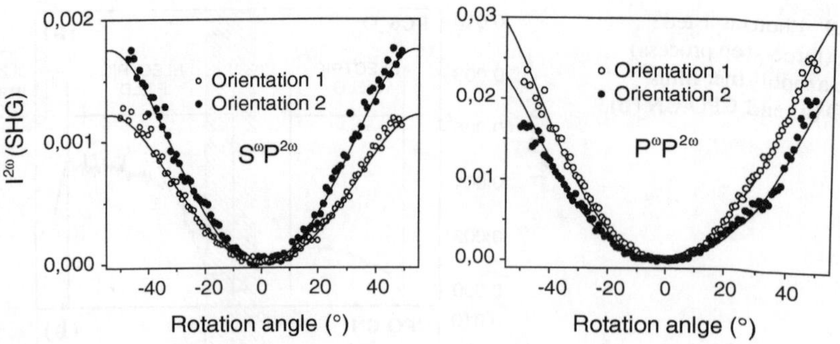

Fig. 5.11 Maker fringes obtained by photoassisted poling for a CPO-CN film

If we measure the refractive indices of the sample oriented by photoassisted poling we find biaxiality in the irradiated area with the following indices measured for the three directions in the case of CPO-CN: $n_x = 1.65$, $n_y = 1.62$ and $n_z = 1.66$. We chose CPO-CN for the following study, because the second harmonic signal obtained from the other two azopolymers is not intense enough. Once it has been verified that the signal does not decay in time, Maker fringes have to be measured for two different orientations of the film (Fig. 5.7) and two polarizations of the incident light, since the sample has C_{2v} symmetry. The fringes thus obtained are shown in Fig. 5.11. Given the difference in intensity between them, the second-order susceptibility tensor is described by three coefficients: $d_{33} = 2.2$ pm/V, $d_{31} = 0.46$ pm/V and $d_{24} = 0.36$ pm/V. We see that for the two orientations of the film the ratio between d_{33} and the corresponding coefficient in the plane, d_{31} or d_{24}, is greater than three, indicating a higher order than that attainable in amorphous polymers under low field conditions. So far, the study is similar to that conducted for azopolymers Pol-PZ-CN in Chap. 4 but in that case the biaxiality of the NLO response obtained was much larger.

Once the behaviour of the matrices has been detailed, the response of the mixtures of azopolymers with Me-3-FX, under the same treatments, will be described. The first step is to check whether the NLO signal of the chromophore dispersed in the matrix is modified by irradiation. For this purpose, samples of the three doped azopolymers were subjected to the same process as the undoped ones. The result is qualitatively similar in all three cases, and it can be seen in Fig. 5.12 for a sample of doped PC6_O.

A clear increase of the second harmonic signal compared to the undoped copolymer is observed, so the chromophore is being oriented under the electric field. In addition, there is a noticeable change in the slope before and after irradiation, as was the case with the films of undoped azopolymers. As can be seen at the bottom of the graph, when the polymer is not irradiated before the second application of the electric discharge, the generated harmonic intensity is almost identical to that achieved during the first poling, so it is proved that the increase of the second harmonic intensity, associated with a more efficient orientation of the chromophores, is caused by

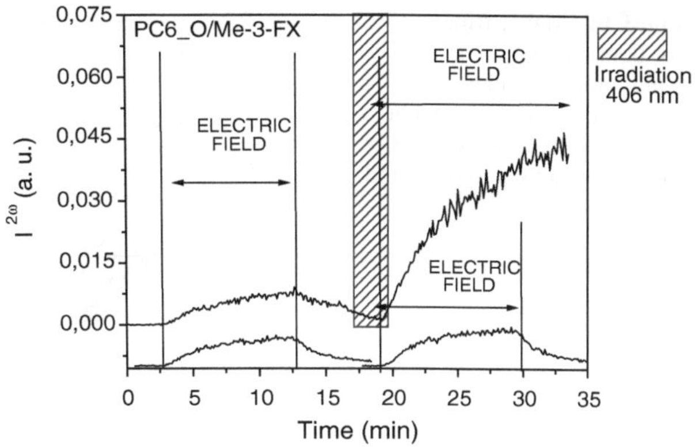

Fig. 5.12 Second harmonic signal intensity for PC6_O/Me-3-FX films as a function of time for the three-step photoassisted poling

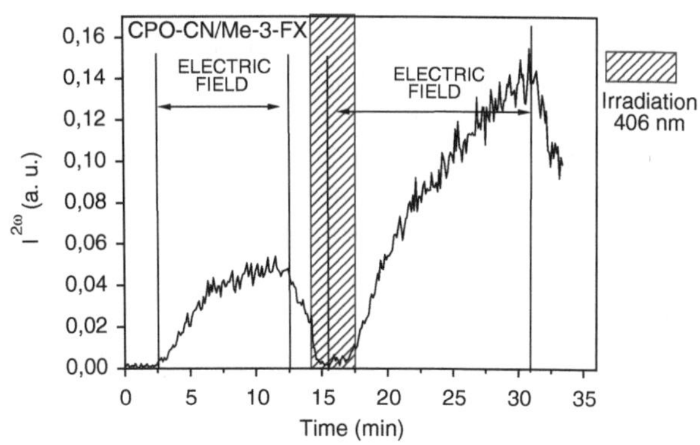

Fig. 5.13 Three-step photoassisted poling for the CPO-CN/Me-3-FX mixture

irradiation. In Fig. 5.13 the same process but for the CPO-CN/Me-3-FX mixture is depicted.

The signal enhancement registered as a result of irradiation depends on several factors, such as the thickness of the film and the intensity of the incident light, and no beneficial effect above a certain maximum intensity is observed.

This increase in the polar orientation of the chromophores by illumination may be due to the enhanced mobility in the matrix related to the presence of *cis* isomers, which generate free volume around the chromophores, thus decreasing the temperature at which the glass transition occurs, and allowing the electric field to orient the donor-π-acceptor moieties.

As shown in Fig. 5.14, the ratio of intensities P^{ω}/S^{ω} is greater for orientation 1 than for orientation 2. However, the low stability of the second harmonic signal thus generated prevents a quantitative measure of the coefficients from being obtained. Thus, thermal poling processes were applied to pre-irradiated samples with an in-plane birefringence of 0.03. PC2_O/Me-3-FX samples maintained this order as shown in Fig. 5.14 and resulted in higher NLO coefficients, $d_{33} = 7$ pm/V, $d_{31} = 1.7$ pm/V and $d_{24} = 1.4$ pm/V, with ratios of $d_{33}/d_{31} = 5$ and $d_{33}/d_{24} = 4$.

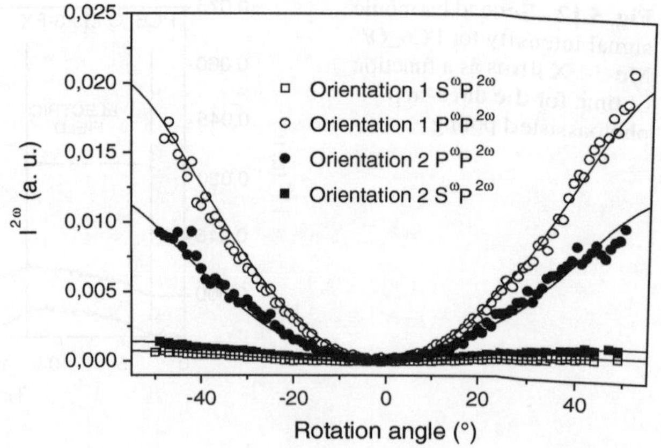

Fig. 5.14 Maker fringes for both measurement configurations and two polarizations of the incident light for a PC2_O/Me-3-FX film oriented by photoassisted poling

However, as in the thermal case, the influence of the order the matrix acquires must be considered in the final orientation of the chromophores, that is, the degree of coupling between them. A stable in-plane birefringence of 0.03 was measured in the films of PC2_O and CPO-CN mixtures, oriented as in Figs. 5.12 and 5.13. PC6_O/Me-3-FX films were discarded because they did not show sufficient optical quality for the evolution of the indices to be tracked and for NLO measurements to be made. The Maker fringes obtained in this way for the two orientations of the films show a different intensity ratio between fringes measured with polarization P^{ω} and S^{ω}.

As the final nonlinear response is very intense, it must necessarily come from the chromophore, and therefore the biaxiality of the response would be forced by the in-plane anisotropy of the matrix.

By contrast, in the case of CPO-CN/Me-3-FX, when it is oriented by thermal poling with previous irradiation, the homeotropic tendency "prevails" and the NLO response is uniaxial, with d_{33}/d_{31} ratios of about 10. Therefore, the idea that the matrix arrangement to some extent imposes the ordering that the dispersed chromophore acquires is reinforced again.

5.4 Conclusions

Two LC azopolymers with different thermotropic tendencies and an amorphous polymer were studied. PC2_O shows in-plane orientation in the film and the PC6_O sample out-of-plane orientation by heating above T_g. However, when these films are oriented by thermal poling, the electric field alters this trend, and the refractive index measurement shows that the polymer remains in a state close to the isotropic initial one. Therefore, the orientation achieved by thermal poling in mixtures of these two azopolymers with a highly efficient chromophore gave rise to $d_{33}/d_{31} = 3$, typical of amorphous materials, which are almost isotropic. It was checked by using the same orientation conditions that for the mixture PMMA (amorphous polymer)/chromophore the mentioned ratio was three.

If the conditions are intensified by raising the temperature of the thermal poling, a slight homeotropy in the matrix is forced, and results in d_{33}/d_{31} ratios around 6. In the case of the CPO-CN copolymer, the electric field disturbs the homeotropic tendency characteristic of this liquid crystal, but by increasing the poling temperature a very high out-of-plane anisotropy is achieved. The copolymer/chromophore mixture gave rise to d_{33}/d_{31} ratios above 10, which highlights that the chromophore is oriented, at least partially, along the direction set up by the matrix.

Likewise, out-of-plane anisotropy was induced prior to thermal poling in PC6_O/Me-3-FX and CPO-CN/Me-3-FX samples, resulting in $d_{33}/d_{31} \approx 6$ for PC6_O and $d_{33}/d_{31} \approx 10$ for the copolymer. It seems therefore that the coupling between chromophore and matrix is larger in the most polar matrix and the importance of the coupling of the two movements is reinforced.

Photoassisted poling has proved to be a useful method for orienting highly efficient nonlinear chromophores in photoaddressable polymer mixtures. The increased mobility associated with the presence of *cis* species in the side chains of the azopolymer makes it easier to orient the chromophore in an electric field. That results in an intense NLO response of the chromophores oriented by light and an electric field. Furthermore, the symmetry of the nonlinear response induced by polarized light is compatible with a polar biaxial orientation, which makes it clear again that the polar orientation of the chromophore is determined by the interactions with the azopolymer matrix.

References

1. A.W. Harper et al., Translating microscopic optical nonlinearity into macroscopic optical nonlinearity: the role of chromophore chromophore electrostatic interactions. J. Opt. Soc. Am. B **15**(1), 329 (1998)
2. J. Reyes-Esqueda et al., Effect of chromophore–chromophore electrostatic interactions in the NLO response of functionalized organic-inorganic sol-gel materials. Optics Communications **198**(1–3), 207 (2001)
3. Z. Sekkat, M. Dumont, Photoinduced orientation of azo dyes in polymeric films. Characterization of molecular angular mobility. Synth. Met. **54**(1–3), 373 (1993)
4. C.W. To, K.Y. Wong, Comparative studies of molecular reorientations in thermal-assisted and photoassisted electric-field poled nonlinear optical polymers. J. Appl. Phys. **100**(7), 073505(1) (2006)
5. E. Ishow et al., A molecular photostirrer for poling non-linear optical chromophores in a polymer matrix. J. Opt. A: Pure Appl. Opt. **4**(6), S197 (2002)
6. B.C. Olbricht et al., Laser-assisted poling of binary chromophore materials. J. Phys. Chem. C **112**(21), 7983 (2008)
7. S.J. Benight et al., Reduced dimensionality in organic electro-optic materials: theory and defined order. J. Phys. Chem. B **114**(37), 11949 (2010)
8. L.R. Dalton et al. in *Organic Thin Films for Photonic Applications, ACS Symposium Series*, vol. 1039, ed. by W.N. Herman, S.R. Flom, S.H. Foulger, ACS, Washington, DC, (2010) (Chap. 2)
9. R.M. Tejedor et al., Photoinduced chiral nematic organization in an achiral glassy nematic azopolymer. Adv. Funct. Mater. **17**(17), 3486 (2007)

10. S. Alías et al., Iminium salts of ω-dithiafulvenylpolyenals: an easy entry to the corresponding aldehydes and doubly proaromatic nonlinear optic-phores. J. Org. Chem. **73**(15), 5890 (2008)
11. J.G. Meier, R. Ruhmann, J. Stumpe, Planar and homeotropic alignment of LC polymers by the combination of photoorientation and self-organization. Macromolecules **33**(3), 843 (2000)
12. M. Han, K. Ichimura, In-plane and tilt reorientation of p-methoxyazobenzene side chains tethered to liquid crystalline polymethacrylates by irradiation with 365 nm light. Macromolecules **34**(1), 90 (2001)
13. E. Uchida et al., Control of thermally enhanced photoinduced reorientation of polymethacrylate films with 4-methoxyazobenzene side groups by irradiating with 365 and 633 nm light and annealing. Macromolecules **37**(14), 5282 (2004)
14. H.W. Guan, C.H. Wang, Dipolar interaction assisted effects on second harmonic generation. J. Chem. Phys. **98**(4), 3463 (1993)
15. A. Vembris, M. Rutkis, E. Laizane, Effect of corona poling and thermo cycling sequence on NLO properties of the guest-host system. Mol. Cryst. Liq. Cryst. **485**, 873 (2008)
16. J.C. Dubois et al., Behavior and properties of side chain thermotropic liquid crystal polymers. Acta Polym. **48**(3), 47 (1997)

Chapter 6
Nonlinear Optical Gratings

This chapter presents the results obtained on recording nonlinear gratings in azopolymers. It should be recalled that a nonlinear grating is a system, a polymer film in our case, in which the polar orientation induced by an electric field has a spatial modulation, resulting in a modulation of the nonlinear susceptibility tensor. When a light beam hits a diffraction grating, a redistribution of the intensity of the transmitted light is observed, a sequence of diffraction maxima in increasing order (± 1, ± 2, …) appearing on both sides of the propagation direction of the incident beam (which determines the direction of the 0 order) as we move away from the incident direction. When a nonlinear grating is recorded in a material, diffraction maxima are observed also in certain directions corresponding to the second harmonic light. In these systems it is possible to produce SHG in "quasi phase matching" conditions. Very efficient SHG can be obtained, given a material in which a modulation of $\chi^{(2)}$ with a certain periodicity ($\wedge$) has been introduced, for the right combination of excitation wavelength/angle of incidence [1, 2].

One of the recording methods is based on the illumination of photo addressable material films with a periodic intensity pattern, which selectively modifies the properties of the material. Chapter 2 briefly described that this variation leads to the formation of diffraction gratings in the material. We have also seen that the irradiation can disorder the material, so if the recording is applied to a polarly oriented film, the deletion can be made in a selective way, leading to the aforementioned modulation of nonlinear susceptibility of the material. On the other hand, if the irradiation (not uniform) of an initially isotropic polymer is carried out at the same time as the application of the electric field, photo assisted poling will occur in the illuminated areas while those receiving no light (or much less light) remain in an almost isotropic state. Again, this leads to a modulation of the nonlinear properties of the material.

Among the materials studied in this thesis, the copolymer with the lower azo content, Copol-PZ-CN_L, was the one chosen to study this phenomenon; its properties are presented in Chap. 4. The motivation for this choice is found in its good nonlinear response, the good reproducibility of the orientation processes, as well

R. Alicante, *Photoinduced Modifications of the Nonlinear Optical Response in Liquid Crystalline Azopolymers*, Springer Theses, DOI: 10.1007/978-3-642-31756-9_6,

Fig. 6.1 Polymers used to record nonlinear gratings

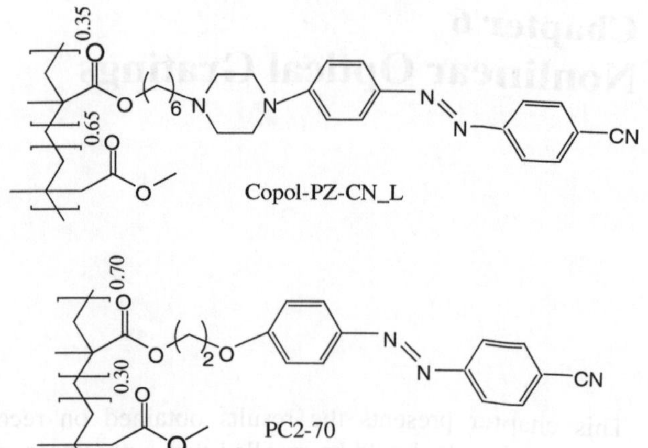

as the efficiency of the photo assisted polar orientation processes. Moreover, the lower incidence of aggregation in this material compared to the homopolymer favours the photo addressing process. Along with this polymer, gratings in an amorphous copolymer, PC2-70, were recorded and characterized. The nonlinear response of PC2-70 has been previously described [3]. The two polymers are shown in Fig. 6.1.

6.1 Recording Procedures

In this study we tested three different methods for the generation of nonlinear gratings. In all of them the modulation of the irradiation intensity was performed by using the interference of two beams overlapping on the film. That results in an intensity distribution such as the one depicted in Fig. 6.2, where a succession of alternating light and dark vertical stripes can be seen. We used two wavelengths, 406(diode laser) and 488 nm (Ar^+ laser), both with horizontal polarization.

The recording of the grating is monitored by measuring the intensity of the first diffraction maximum, so that, for example, the effect of applying the field can be seen (see Fig. 2.27).

1. The first process consists in first orienting the polymer using conventional thermal poling and then performing a selective erase of the polar order with light. Once it has been checked that the poling has been effective by measuring the Maker fringes, the films are illuminated with the mentioned interference pattern.

2. A second method is based on applying the electric field so that it coincides during a certain time interval with the irradiation. The polar orientation is therefore achieved by photo assisted poling in the illuminated areas.

Fig. 6.2 Light intensity distribution when two beams interfere

3. Finally, an electric field was applied to films with previously recorded gratings, in which the generated surface relief was significant. Generally, in this process it is necessary to raise the temperature to make the polar orientation effective, unless the field is applied immediately after the illumination.

As regards the first method, preliminary tests have not led to the expected results. Two extreme conditions were tested. Selective illumination of polarly oriented areas for short times ($\approx$1 min) resulted in the formation of inefficient gratings, though the nonlinear response in the zone was maintained. In contrast, when the gratings were recorded for long times (>15 min), the efficiency was right but the nonlinear response decreased dramatically. This may be due to a lack of stability of the interference pattern generated on the film, which causes a loss of contrast in the illumination, so that for long times it becomes practically uniform. Alternatively, if the intensity of the two beams is not exactly equal, the interference minima are nonzero and that light, although less intense, can cause the deletion.

Concerning the second method proposed, several gratings were recorded successfully in both polymers at room temperature with light of 406 and 488 nm wavelength. However, the characterization of the nonlinearity of these gratings is more complicated, owing to the lower stability of the polar orientation achieved by photo assisted poling, as discussed in Chap. 6. It was decided therefore, that it would be more convenient to begin this study, which included the preparation and optimization of the experimental setup, with a recording method that would produce a more stable signal, as is the case for poling at high temperature.

6.2 Results

Let us focus, then, on the process of preparation and characterization of the gratings produced by the third method, when the thermal poling is made on the previously recorded grating.

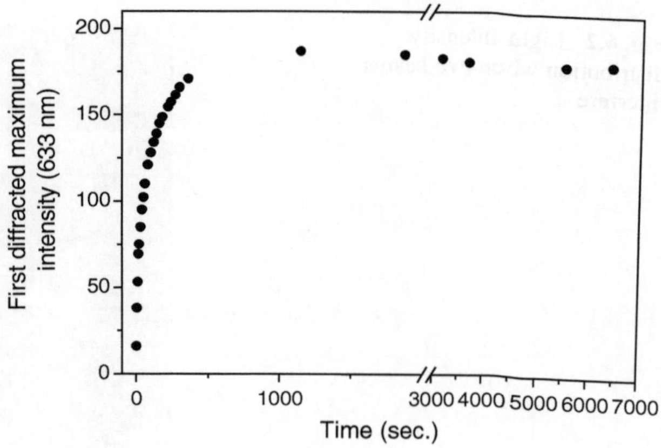

Fig. 6.3 Evolution of the diffracted intensity during a grating recording in a PC2-70 film at 406 nm

6.2.1 Recording of Gratings

Several gratings were recorded using both the mentioned wavelengths in Copol-PZ-CN_L, a LC copolymer with high NLO response, and in PC2_70, which is an amorphous polymer. We should bear in mind that the wavelength and the angle between the beams (2θ) (see Fig. 2.26) that interfere on the film determines the periodicity of the grating (Λ):

$$\Lambda = \lambda/2\ (\sin\theta) \tag{6.1}$$

In this study, the value of θ was between 2.5 and 2°, which fixes the value of the distance between the fringes in the interval 5 μm < Λ < 6.5 μm.

Figure 6.3 shows the evolution of the first-order diffraction maximum intensity corresponding to a recording of a grating in PC2-70. The recording was performed with light of 406 nm wavelength, with an approximate intensity of 50 mW/cm^2 in each beam. The film thickness was close to 2 μm.

It can be seen that after a first stage in which the diffraction efficiency of the grating increases quickly, a level is reached after about 10–15 min that remains essentially stable (except for the fluctuations associated with the already mentioned difficulty of maintaining the contrast between the fringes for such long recording times).

During the recording process, the azochromophores orient selectively in the illuminated zones, and a surface relief modulation with the same periodicity as the grating is produced (see Fig. 6.4). The magnitude of the relief depends on various factors, such as the intensity of the irradiation, the film thickness and the distance between fringes. For the recording shown in Fig. 6.3 the relief reached 100 nm, being relatively uniform in the centre of the two beams' overlap zone.

Several 15 min recordings in various films of similar thickness gave reproducible results for the relief amplitude and the approximate diffraction efficiency (accurate quantitative measurements were not made). The gratings recorded under these

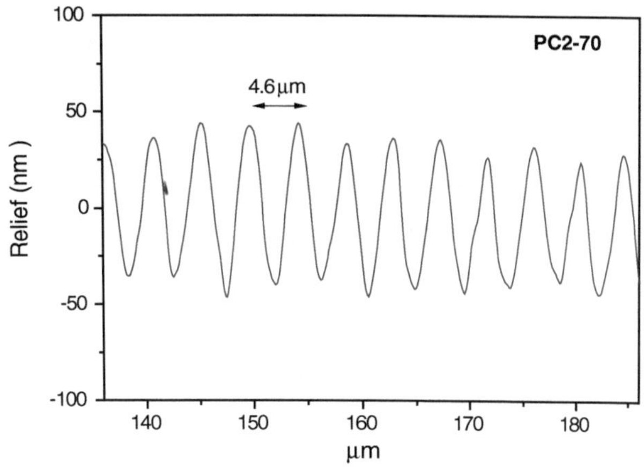

Fig. 6.4 Relief measured in a PC2-70 film after the recording process of Fig. 6.3

conditions in the two materials studied in this section were subsequently oriented by corona poling. In the following section the results obtained taking into account the different nonlinear responses of the materials (clearly greater for the polymer with piperazine) will be compared. The effect of the different depth of the relief on the diffraction efficiency of the nonlinear grating will also be discussed.

6.2.2 *Recording of Nonlinear Gratings*

Once the gratings were recorded, we proceeded to electric field orientation by applying a corona discharge using the method described in Chap. 2 (from RT to T_p at 5 °C/min, 40 min at T_p, and subsequent cooling with the electric field on). In the case of PC2-70, T_p was 125 °C, and for Copol-PZ-CN_L, 105 °C. In all the cases Maker fringes were measured in the grating zone to characterize the nonlinear response of each film. Since the excitation beam size ($\lambda = 1.9$ μm) was larger than the grating periodicity, the second harmonic intensity detected in these measurements corresponded to the average of the illuminated and non-illuminated areas. In these measurements, the nonlinear response was much greater in the case of the piperazine copolymer, and the estimated coefficients were comparable to those obtained in the measurements of this same polymer after conventional thermal poling (see Chap. 4). Also, in these films the relationship between the coefficients d_{33} and d_{31} corresponds to a high anisotropy, while the d_{33}/d_{31} ratio measured for PC2-70 was three, as expected when dealing with an amorphous polymer.

After the orientation process with the electric field, and once it had been verified that the poling had been effective, the surface relief was measured again. It was found that there were no appreciable alterations in the grating zone in PC2-70, and in some films of Copol-PZ-CN_L an increase in the depth of the modulation was observed, when the starting relief was marked enough.

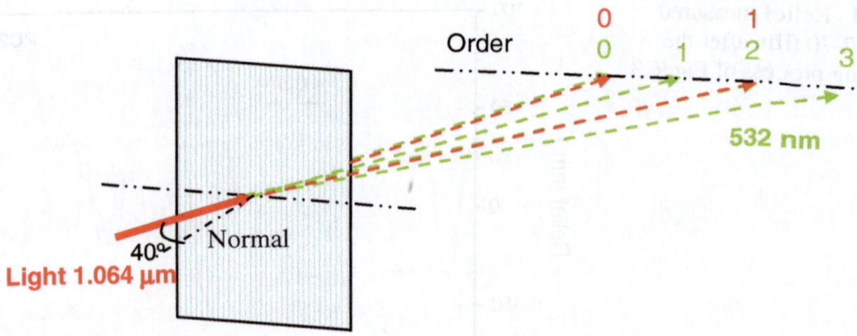

Fig. 6.5 Schematic representation of the diffraction maxima directions of the fundamental beam and the second harmonic

The observation of the second harmonic diffraction maxima generated in the grating zone evidences the modulation of the nonlinearity. This was carried out with a 1.064 μm beam in an oblique direction, forming an angle of about 40° with the normal. The beam was focused on the grating by a lens of ≈30 cm focal length. Our goal was to detect light of 532 nm wavelength in directions that are different from that of the 1.064 μm wavelength excitation beam and perfectly defined by the periodicity of the grating. The diffraction maxima of the fundamental light are easily detectable by a photosensitive card, and the position of the maxima for the harmonic light is such that the first 532 nm maximum is between the direct beam and the first diffraction peak of 1.064 μm; the second order maximum overlaps with the first maximum of 1.064 μm and so on, as shown in Fig. 6.5. It was possible to isolate the light of 532 nm corresponding to the different maxima by using a photomultiplier fitted with suitable interference filters.

6.2.2.1 Gratings with Large Relief in PC2-70 and Copol-PZ-CN-L

The intensity of the diffracted second harmonic in gratings recorded in thick films (≈2 μm) has been compared. The maximum relief of the analyzed gratings was approximately 100 nm, as seen in Fig. 6.6, which corresponds to a grating recorded by the interference of two beams of 488 nm wavelength (≈40 mW/cm^2 each beam).

The piperazine copolymer presents a slight absorption at 532 nm, implying that part of the second harmonic light generated does not "leave" the sample. Despite this, the intensity of the harmonic light detected in the direction of the incident beam of 1.064 μm, at about 40° from the normal as we said (0 order in Fig. 6.5), was clearly greater than that detected for PC2-70, because the polymer with piperazine has a much larger nonlinear response. For a first characterization of the process efficiency, we calculated the ratio of the intensity detected in the second harmonic light first-order peak and in the 0 order (incident beam direction) at 40° of incidence. These ratio values were approximately 0.1 and 0.3 for PC2-70 and Copol-PZ-CN_L, respectively.

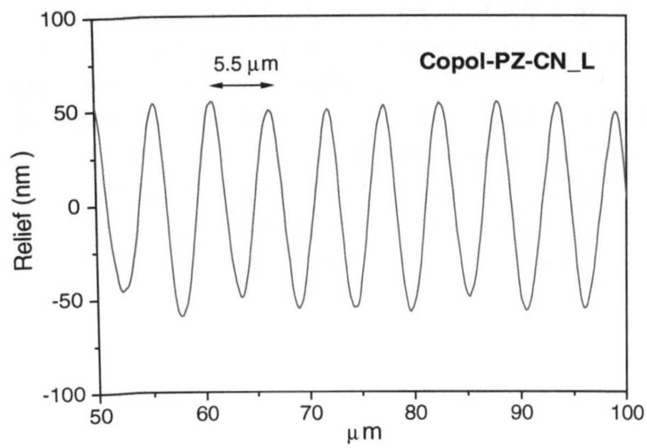

Fig. 6.6 Relief measured in a Copol-PZ-CN-L film after the nonlinear grating recording process

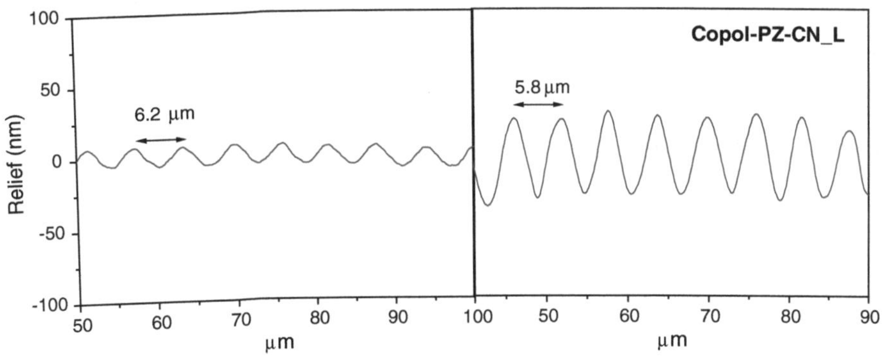

Fig. 6.7 Surface relief in Copol-PZ-CN-L films

6.2.2.2 Influence of Relief on the Grating Efficiency

In order to analyze the influence of the magnitude of the relief in the second harmonic light diffraction, gratings were recorded with light of 406 nm wavelength in Copol-PZ-CN-L films (of equal thickness) under similar conditions except for the time of irradiation. Thus, we obtained films with the relief shown in Fig. 6.7.

In this case, the intensity of the second harmonic detected in the incident beam direction (for 40° incidence) was virtually the same, as was expected for the same material oriented in analogous conditions. However, the diffraction efficiency was clearly higher (more than an order of magnitude) in the case of the film with the larger relief.

Finally, it should be noted that for the recording conditions used in this work, it was easier to obtain high relief in thicker films (2 μm vs. <1 μm). The greatest diffraction efficiency found was 30 % in films of Copol-PZ-CN_L with relief exceeding 100 nm, such as the one depicted in Fig. 6.6.

As commented, in order to get a very intense second harmonic signal in a well-determined direction, it is necessary to produce coupling between the fundamental and the harmonic light in the grating, when illuminating the nonlinear grating [1, 2]. That condition is fulfilled more easily for small Λ and film thicknesses under 1 μm. In our experimental setup, the control of the distance between fringes is relatively simple. Gratings with half of the periodicity of those described so far have been recorded in films with thicknesses greater than 1 μm, with a well-defined relief. However, the lack of reproducibility of relief formation in films of ≈0.5 μm obtained so far will require greater control of all the process conditions, including the preparation of the films.

6.3 Conclusions

Diffraction grating recording was achieved in films of amorphous and LC azopolymers with different relief using light of two different wavelengths. It has been proved that the application of a corona discharge at high temperature induces NLO properties in such gratings.

Diffraction efficiency of the second harmonic light generated in the grating zone has been correlated with the amplitude of the modulation of the surface relief.

References

1. Y. Che et al., Stable surface relief grating with second-order nonlinearity on urethane-urea copolymer film. Opt. Mater. **21**(1–3), 79 (2003)
2. F. Lagugné-Labarthet et al., Significant enhancement of the optical second harmonic generation in a poled azopolymer thin grating. J. Phys. Chem. B **110**(28), 13689 (2006)
3. F.J. Rodríguez et al., Optical anisotropy and non-linear optical properties of azobenzene methacrylic polymers. Polymer **45**(7), 2341 (2004)

Chapter 7
General Conclusions

In order to close this report, the conclusions reached during its development will be summarized below.

Regarding the analysis made in Chap. 3 on the molecular nonlinear optical response, it has been proven that the two-level model is suitable for the calculation of nonlinear static hyperpolarizabilities in the case of dipolar azo compounds with rod-type structure. The study of the NLO properties of merocyanines by EFISH confirmed that the proaromatic character of both the giver and acceptor group enhances the hyperpolarizability value.

A comprehensive study of the effect of changing the length, morphology and structure of the spacer chain on the NLO response was performed. The cycling of the spacer chain did not result in an improvement of the NLO properties in all cases, revealing that the response does not depend only on the structure of the spacer. On the one hand, spacer groups containing pyran improved the compromise between good stability and high response, compared to the isophorone group. On the other hand, it was proven that the lengthening of the spacer with a heterocycle on the acceptor group's side improves significantly the NLO response, compared to the lengthening on the donor group's side. Moreover, we also studied the change of the sign of the hyperpolarizability value that different families of compounds experience when the spacer length, the donor or acceptor group type, or the solvent is varied. This behaviour is associated with different regions of Marder's curve. The best NLO response, at the molecular level, was obtained for the compounds with diphenylpyran as the donor group and tricyanofuran substituted with trifluoromethylene as the acceptor one.

The results obtained from the EFISH measurements were decisive for selecting chromophores that maintain a good balance between nonlinearity on the one hand, and stability and transparency on the other, with the aim of their incorporation into polymers.

The study about side-chain LC azopolymers, reported in Chap. 4, led to the conclusions listed below:

The LC azopolymer films studied show a strong homeotropic tendency. The axial order reaches very high and stable values because of the thermotropic effect.

R. Alicante, *Photoinduced Modifications of the Nonlinear Optical Response in Liquid Crystalline Azopolymers*, Springer Theses, DOI: 10.1007/978-3-642-31756-9_7,

It proved possible to modify the anisotropy in Pol-PZ-CN films by changing the strong homeotropic tendency into a planar distribution of the chromophores by means of optical and thermal treatments.

The presence of azobenzene aggregates in the Pol-PZ-CN homopolymer films was confirmed by studying the absorption spectra under thermal treatment. These H-type aggregates are oriented preferably perpendicular to the plane of the film and show lower oscillator strength than the isolated species. The thermotropic effect clearly favours the formation of these aggregates. Irradiation with UV-blue light partially destroys this aggregation, but the aggregated initial state is recovered when the film is re-dissolved and reconstituted.

Concerning the NLO properties, the introduction of a strong donor group in the side chain of the LCP resulted in an efficiency value below that estimated by theoretical calculations by assuming the additivity of the microscopic properties. It was evidenced that this behaviour is related to the presence of aggregates, which are much more significant in the films of polymers bearing chromophores with higher dipole moment. However, the nonlinear response remains stable for a long time.

It was found that during thermal poling processes there is competition between the polar orientation induced by the electric field and the antiparallel orientation (centrosymmetric) established by the thermotropic effect, which does give rise to a NLO response. The pre-irradiation with UV-blue light provides access to more isolated azochromophores during the thermal poling, so that we obtain larger nonlinear coefficients values without diminishing their stability.

Pol-PZ-CN film orientation was achieved by photo assisted poling at low temperature. The stability of the induced polar order increases with the temperature at which the process is performed.

The d_{33}/d_{31} ratios found are much higher than three, indicating a high order, with a distribution around the polar axis narrower than that typically obtained in amorphous polymers. Furthermore, in processes that include blue light irradiation an increase in this ratio was obtained.

The modification of the second-order susceptibility tensor symmetry with respect to that usually obtained after thermal poling was achieved by means of optical and thermal treatments. The tensor symmetry changes from $C_{\infty v}$ to a biaxial nonlinear response described by a C_{2v} tensor.

A decrease in the azochromophore content of the copolymers based on mon-PZ-CN led to large and stable NLO responses (higher than those of the homopolymer). The d_{33}/d_{31} ratios were lower, corresponding to a wider distribution of oriented dipoles. These results are associated with the reduction of aggregation that enables the orientation of a larger number of isolated chromophores and increases the free volume in the film.

The study of the low or weakly polar azopolymer mixtures with dispersed highly efficient nonlinear chromophores described in Chap. 5 showed the existence of a motion coupling of the chromophores and the side chains of the polymeric matrix. During thermal poling processes, the electric field disturbs the orientation of the thermotropic liquid crystals with an almost symmetrical side chain. Very high NLO coefficients are obtained, assignable to the oriented chromophore, but ratios $d_{33}/d_{31} = 3$, typical of amorphous materials. However, when the dipole moment of the side chain is increased, high values of anisotropy are reached. The NLO response, largely due to the dispersed chromophores, presents ratios well above three. Furthermore, the thermal induction of out-of-plane anisotropy in the polymeric matrix prior to the thermal poling process also resulted in d_{33}/d_{31} ratios typical of liquid crystals. Finally, it was found that this coupling also occurs in the orientation processes using UV-blue light, since doped azopolymer films with in-plane anisotropy induced by light have a response, after thermal poling, whose intensity is assignable to the chromophore and that is described by a susceptibility tensor with C_{2v} symmetry.

In Chap. 6 it was proved that the application of a corona discharge to a diffraction grating that was previously recorded with light in an azopolymer results in a nonlinear grating owing to the modulation of the susceptibility tensor of the polymer. Second harmonic light was detected in well-defined directions, different from the direction of incident light, relating its intensity to the amplitude of the surface relief.

Appendix A
Expression for the Second Harmonic Intensity at the Exit of the Nonlinear Planar Surface

Maxwell's equations are the fundamental equations of electromagnetism and the basis for describing electromagnetic interactions in any medium (usually not homogeneous, nonlinear and isotropic). In their differential form (when written in the International System of Units, SI) they can be expressed in terms of **E** and **B** vectors, the intensity of the electric field and the magnetic induction, respectively:

$$\nabla \cdot \vec{E} = \frac{\rho_t}{\varepsilon_0} \tag{A.1}$$

$$\nabla \times \vec{E} = -\frac{\partial \vec{B}}{\partial t} \tag{A.2}$$

$$\nabla \cdot \vec{B} = 0 \tag{A.3}$$

$$\nabla \times \vec{B} = \frac{1}{c^2}\frac{\partial \vec{E}}{\partial t} + \mu_0 \vec{J_m} \tag{A.4}$$

where ρ_t is the total charge density (both free and bound) and $\boldsymbol{J}_m$ is the current density due to movement of charges in the material. In addition, we have the constitutive equations, characteristic of each particular medium. Thus, in a linear and isotropic medium, the relationship between the **E**, **D** and **P** vectors (electric field, electric displacement and polarization of the medium, respectively) is given by

$$\vec{D} = \varepsilon_0\vec{E} + \vec{P} = \varepsilon_0\varepsilon_r\vec{E}, \tag{A.5}$$

where ε_0 is the vacuum permittivity and ε_r the relative permittivity or dielectric constant of the medium.

The relationship between **B** and the magnetic field **H** and the magnetization of the medium **M** is

$$\vec{B} = \mu_0(\vec{H} + \vec{M}) = \mu_0\mu_r\vec{H}, \tag{A.6}$$

where μ_0 is the vacuum magnetic permittivity and μ_r the relative permittivity of the medium.

R. Alicante, *Photoinduced Modifications of the Nonlinear Optical Response in Liquid Crystalline Azopolymers*, Springer Theses, DOI: 10.1007/978-3-642-31756-9,

We will study the propagation of light through non-magnetic dielectric media. Therefore, $\boldsymbol{B} = \mu_0 \boldsymbol{H}$, and the only contribution to the current intensity is the one associated with the polarization, $\boldsymbol{J}_{\mathrm{m}} = \partial \boldsymbol{P}/\partial t$. When these relations are introduced into the Faraday (A.2) and Ampère–Maxwell equations (A.4), these are expressed as

$$\nabla \times \vec{E} = -\mu_0 \frac{\partial \vec{H}}{\partial t}, \tag{A.7}$$

$$\nabla \times \vec{H} = \varepsilon_0 \frac{\partial \vec{E}}{\partial t} + \frac{\partial \vec{P}}{\partial t} \tag{A.8}$$

Taking the curl in A.7 and replacing the result with A.8, we have[1]

$$\nabla \times \nabla \times \vec{E} = -\varepsilon_0 \mu_0 \left(\frac{\partial^2 \vec{E}}{\partial t^2} \right) - \mu_0 \left(\frac{\partial^2 \vec{P}}{\partial t^2} \right), \tag{A.9}$$

which is the propagation equation of electromagnetic waves in a medium.

In an anisotropic medium, ε_{r} is not a scalar but a tensor. In the case of a nonlinear medium, Eq. A.5 must include a polarization term that is nonlinear with the field ($\propto E^2, E^3, \ldots$), which also appears in the wave propagation equation.

Previous Considerations

The aim is to obtain the second harmonic intensity expression (proportional to the square of the amplitude of the harmonic frequency field) at the output of a thin film. Figure A.1 shows a scheme of the problem and some variables that will appear in development are defined.

We consider an electromagnetic wave of fundamental frequency ω hitting a thin layer of thickness L at an angle θ. E_ω is the electric field amplitude and $\boldsymbol{k}_\omega$ the wave vector of the incident beam, perpendicular to the Y-axis (around which the sample will rotate). The refracted wave of frequency ω will propagate inside the layer in the direction given by Snell's law and its amplitude E_ω' will depend on the air/sample transmission factor $\left| E_\omega' \right| = t_\omega' \left| E_\omega \right|$.

The electric field E_ω' inside induces a polarization in the medium, which is the sum of a linear term (oscillating at the fundamental frequency) and various nonlinear ones, among which only the second-order (2ω oscillating frequency) will be of interest to us.

This polarization radiates electromagnetic waves of angular frequency 2ω, called "bound" harmonic waves. In addition, there will also be "free" harmonic waves with frequency 2ω generated on the surface. The two waves will propagate at different rates, n^ω and $n^{2\omega}$ being the refractive indices of the bound and free

[1] $\nabla \times \nabla \times \vec{E} = -\frac{\varepsilon_r}{c^2} \left(\frac{\partial^2 \vec{E}}{\partial t^2} \right) - \frac{4\pi}{c^2} \left(\frac{\partial^2 \vec{P}}{\partial t^2} \right)$ in the c.g.s. system

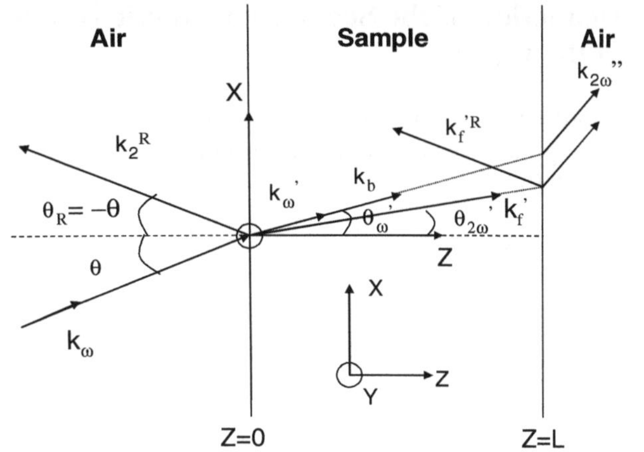

Fig. A.1 Propagation of waves of frequency ω and 2ω through a thin layer of thickness L. k_ω = fundamental wave, k_b = bound harmonic wave, k_f' = free harmonic wave, k_2^R and $k_f'^R$ = reflected waves. The primes and double primes refer to the values inside the sample (') and in the air (″), respectively, after crossing the layer

waves, respectively. When the sample is rotated around Y, the light path will vary and, as a consequence of the interference of the two waves, a second harmonic intensity $I_{2\omega}''$ will be observed that oscillates with the sample thickness passed through, i.e. with the angle of rotation.

In order to obtain the expression for $I_{2\omega}''$ generated by the sample as a function of angle θ, the work of Jerphagnon and Kurtz, (hereafter JK) [1] will be followed. During the development of their work they made several assumptions.

1. The sample does not absorb either at the fundamental or at the second harmonic frequency. This condition was verified in all the polymers studied in this thesis.
2. The Poynting vector is taken to be collinear with the harmonic and fundamental wave vectors, i.e. it is not taken into account that the free and bound waves do not travel exactly in the same direction inside the crystal. This assumption is acceptable in isotropic materials with low dispersion. In the case of the LCPs studied in this work it is necessary to introduce a correction to this model due to their considerable [2] anisotropy. In 1994 Herman and Hayden proposed a suitable model for birefringent samples (which in addition takes into account dispersion and absorption effects). The details of the correction will be given at the end of this appendix.
3. It is assumed that there are no multiple reflections on the input and output surfaces. This is a good approximation for low refractive index samples, suitable for the studied polymeric films.
4. It is also assumed that the transverse dimension of the sample is infinite and there are no effects due to the beam size, in whose cross section (A) it is considered that the field amplitude is constant (and $A >> L^2$). If the beam size is too large some corrections will need to be introduced. These corrections will be important only for high θ values, in the case of very thick and dispersive films or highly focused light.

Deduction of the Second Harmonic Intensity at the Exit of the Layer

In order to calculate the electric field $E_{2\omega}'$ at each point of the layer, a term representing the existence of nonlinear polarization, $P_{2\omega}'$, has to be included in the wave Eq. (A.9):

$$\vec{P}_{2\omega}'(\vec{r},t) = \hat{p}P_{2\omega}'(\vec{r},t) = \frac{1}{2}\left[\vec{P}_{2\omega}' \exp(i\vec{k}\cdot\vec{r})\exp(-i\omega t) + c.c.\right]. \quad \text{(A.10)}$$

Here $\hat{p}$ is a unit vector in the direction of the nonlinear polarization and $P_{2\omega}'$ is the amplitude of the polarization (time-independent), whose relationship with the electric field is given by

$$\vec{P}_{2\omega}' = \chi^{(2)}(-2\omega;\omega,\omega)\vec{E}_{\omega}'\vec{E}_{\omega}'$$

By introducing the polarization of Eq. A.10 in Eq. A.9, we have

$$\nabla\times\nabla\times\vec{E}_{2\omega}'(\vec{r},t) + \left(\frac{(n^{2\omega})^2}{c^2}\right)\left(\frac{\partial^2\vec{E}_{2\omega}'(\vec{r},t)}{\partial t^2}\right) = -\left(\frac{4\pi}{c^2}\right)\left[\frac{\partial^2\vec{P}_{2\omega}'(\vec{r},t)}{\partial t^2}\right]. \quad \text{(A.11)}$$

The solution to this differential equation can be written as a sum of two terms:

$$\underbrace{\underbrace{\vec{E}_{2\omega}' = \hat{e}_f E_f' \exp(i\vec{k}_f\cdot\vec{r})}_{\text{Free harmonic}} + \left[4\pi P_{2\omega}'/\left((n^{\omega})^2 - \left(n^{2\omega}\right)^2\right)\right]\times\left[\hat{p} - \left(\vec{k}_b(\vec{k}_b\cdot\hat{p})/\left|k_f\right|^2\right)\right]\exp(i\vec{k}_b\cdot\vec{r})}_{\text{Bound harmonic}} \quad \text{(A.12)}$$

where $\hat{e}_f$ is a unit vector perpendicular to the free wave propagation; $\vec{k}_f$ and $\vec{k}_b$ are the propagation vectors of the free and bound waves respectively. In order to simplify the mathematical expressions in the following development we will define

$$\hat{e}_b = \hat{p} - \left[\vec{k}_b(\vec{k}_b\cdot\hat{p})/\left|k_f\right|^2\right], \quad \text{(A.13)}$$

the unit vector in the direction of the electric field associated with the bound wave. Also

$$Q' = 4\pi P_{2\omega}'/\left((n^{\omega})^2 - \left(n^{2\omega}\right)^2\right) \quad \text{(A.14)}$$

will be the amplitude of the bound wave.
The electric field will be

$$\vec{H}_{2\omega}' = (c/2\omega)(\vec{k}_f\times\hat{e}_f)E_f'\exp(i\vec{k}_b\cdot\vec{r}) + \left[4\pi P_{2\omega}'\Big/\left((n^{\omega})^2 - (n^{2\omega})^2\right)\right]\cdot(c/2\omega)(\vec{k}_b\times\hat{p})\exp(i\vec{k}_b\cdot\vec{r}) \quad \text{(A.15)}$$

In the calculation of the second harmonic intensity at the exit, we will consider several steps, taking into account the boundary conditions at $Z = 0$ and $Z = L$, and the generation and propagation of harmonic waves in the material.

Boundary Conditions at the Input Surface of the Sample ($Z = 0$)

The fundamental wave arriving at an angle θ to the surface at $Z = 0$, is refracted at an angle given by Snell's law.

$$\sin\theta'_{\omega} = \frac{\sin\theta}{n^{\omega}}. \tag{A.16}$$

The propagation vector modulus will be

$$\left|\vec{k}_{\omega}\right| = n^{\omega}\left(\frac{\omega}{c}\right), \tag{A.17}$$

which will be the wave vector. Similarly, for the free harmonic waves we have

$$\sin\theta'_{2\omega} = \sin\theta/n^{2\omega} \text{ and } \left|\vec{k}_{f}\right| = n^{2\omega}\left(\frac{2\omega}{c}\right) \tag{A.18}$$

and for the bound waves

$$\sin\theta'_{\omega} = \sin\theta/n^{\omega} \text{ and } \left|\vec{k}_{b}\right| = n^{\omega}\left(\frac{2\omega}{c}\right). \tag{A.19}$$

There also exists in the air a wave of frequency 2ω reflected at the input surface of the sample, which propagates at an angle $\theta_R = -\theta$, given by

$$\vec{E}^{R}_{2\omega} = \hat{e}_R E'_R \exp(i\vec{k}^{R}_{2}\cdot\vec{r}), \tag{A.20}$$

$$\vec{H}^{R}_{2\omega} = (c/2\omega)(\vec{k}^{R}_{2}\times\hat{e}_R)E_R \exp(i\vec{k}^{R}_{2}\cdot\vec{r}), \tag{A.21}$$

where $\left|k^{R}_{2}\right| = 2\omega/c$.

The values of E_f', E_R, $\hat{\boldsymbol{e}}_f$ and $\hat{\boldsymbol{e}}_R$ need to be found by imposing the boundary conditions, that is, the continuity of the tangential components of the electric and magnetic fields.

By introducing Eqs. A.13 and A.14 into A.12 and A.15, and since $r = 0$ at the entry of the sample, we obtain the following expressions for the electric and magnetic fields at the input to the sample:

$$\vec{E}'_{2\omega} = \hat{e}_f E'_f + \hat{e}_b Q', \tag{A.22}$$

$$\vec{H}'_{2\omega} = (c/2\omega)[(\vec{k}_f\times\hat{e}_f)E'_f + Q'(\vec{k}_b\times\hat{p})]. \tag{A.23}$$

The mentioned continuity relations concern the tangential components, X and Y of the electric and magnetic fields, in this case. Therefore, by setting equal the waves of frequency 2ω before and after the boundary surface:

$$\hat{X}.\hat{e}_R E_R = \hat{X}.\hat{e}_f E'_f + \hat{X}.\hat{e}_b Q', \tag{A.24}$$

$$(\vec{k}_R \times \hat{e}_R).\hat{X} E_R = \hat{X}.(\vec{k}_f \times \hat{e}_f) E'_f + \hat{X}.(\vec{k}_b \times \hat{p}) Q', \tag{A.25}$$

$$\hat{Y}.\hat{e}_R E_R = \hat{Y}.\hat{e}_f E'_f + \hat{Y}.\hat{e}_b Q', \tag{A.26}$$

$$(\vec{k}_R \times \hat{e}_R).\hat{Y} E_R = \hat{Y}.(\vec{k}_f \times \hat{e}_f) E'_f + \hat{Y}.(\vec{k}_b \times \hat{p}) Q'. \tag{A.27}$$

In order to simplify the notation, we will name the projections of the $\hat{e}_b$, $\hat{e}_f$, $\hat{e}_R$ and $\hat{p}$ unit vectors in X, Y, Z, as follows:

	$\hat{e}_b$	$\hat{e}_f$	$\hat{e}_R$	$\hat{p}$
X	b_x	f_x	R_x	p_x
Y	b_y	f_y	R_y	p_y
Z	b_z	f_z	R_z	p_z

By developing the scalar and vector products and remembering that $k_{Ry} = k_{fy} = k_{by} = 0$, as the XZ is the plane of incidence, we find the four equations

$$R_x E_R = f_x E'_f + b_x Q', \tag{A.28}$$

$$R_y k_{Rz} E_R = f_y k_{fz} E'_f + b_y k_{bz} Q', \tag{A.29}$$

$$R_y E_R = f_y E'_f + p_y Q', \tag{A.30}$$

$$(R_x k_{Rz} - R_Z k_{Rx}) E_R = (f_x k_{fz} - f_z k_{fz}) E'_f + (p_x k_{bz} - p_z k_{bx}) Q'. \tag{A.31}$$

We are going to consider, on the one hand, the situations in which the nonlinear polarization is perpendicular and, on the other, when it is parallel to the plane of incidence, XZ.

Nonlinear Polarization Perpendicular to the Plane of Incidence (XZ): $b_x = b_z = 0$, $b_y = 1$; $p_x = p_z = 0$, $p_y = 1$

In this case, by particularizing Eqs. A.28–A.31,

$$R_x E_R = f_x E'_f, \tag{A.32}$$

$$R_y k_{Rz} E_R = f_y k_{fz} E'_f + k_{bz} Q', \tag{A.33}$$

$$R_y E_R = f_y E'_f + Q', \tag{A.34}$$

$$(R_x k_{Rz} - R_Z k_{Rx}) E_R = (f_x k_{fz} - f_z k_{fz}) E'_f. \tag{A.35}$$

Equations A.32 and A.35 are independent of Q′ (bound wave amplitude), and would be equal for the case where there is no second order polarization, that is, if there were no field at 2ω. It follows that $\boldsymbol{R_x} = \boldsymbol{f_x} = 0 = \boldsymbol{R_z} = \boldsymbol{f}$ and $\boldsymbol{f_y} = \boldsymbol{R_y} = 1$, so that Eqs. A.34 and A.35 are now

$$k_{Rz} E_R = k_{fz} E_f^{'} + k_{bz} Q^{'}, \tag{A.36}$$

$$E_R = E_f^{'} + Q^{'}. \tag{A.37}$$

In A.36 we can replace the projections along the Z-axis of the free, bound and reflected wave vectors by

$$k_{fz} = \cos\theta_{2\omega}^{'} \left|k_f\right|, \; k_{Rz} = -\cos\theta \left|k_R\right|, \; k_{bz} = -\cos\theta_{\omega}^{'} \left|k_b\right|,$$

giving, from Eq. A.37,

$$n^{2\omega} \cos\theta_{2\omega}' E_f' + n^{\omega} \cos\theta_{\omega}' Q' = -\cos\theta E_R. \tag{A.38}$$

In turn, by replacing E_R in this equation by its value in A.37, we obtain an expression for the amplitude of the free second harmonic wave inside the sample:

$$E_f^{'} = -Q^{'} \frac{\cos\theta + n^{\omega} \cos\theta_{\omega}^{'}}{\cos\theta + n^{2\omega} \cos\theta_{2\omega}^{'}}. \tag{A.39}$$

Going back now to the expression for the electric field A.12, we can rewrite it as

$$\boxed{\vec{E}_{2\omega}^{'} = Q^{'}\,\hat{Y}\left(\left(-\frac{\cos\theta + n^{\omega}\cos\theta_{\omega}^{'}}{\cos\theta + n^{2\omega}\cos\theta_{2\omega}^{'}}\right)\exp(i\vec{k}_f\cdot\vec{r}) + (\exp(i\vec{k}_b\cdot\vec{r})\right).} \tag{A.40}$$

Nonlinear Polarization Parallel to the Plane of Incidence (XZ): $p_y = 0$, $b_y = 0$

From Eq. A.13, the components in each direction of the polarization unit vector for the case in hand can be deduced:

$$b_x = p_x - \left[k_{bx}(\vec{k}_b \cdot \hat{p}) / \left|k_f\right|^2\right]. \tag{A.41}$$

Developing the dot product, and in view of the values of the components of the wave vectors, we find

$$b_x = p_x - \left(\left(n^{\omega}\right)^2 / \left(n^{2\omega}\right)^2\right) \sin\theta_{\omega}^{'}(\sin\theta_{\omega}^{'} p_x + \cos\theta_{\omega}^{'} p_z). \tag{A.42}$$

Similarly, for the component in the Z direction we will have

$$b_z = p_z - \left(\left(n^{\omega}\right)^2 / \left(n^{2\omega}\right)^2\right) \cos\theta_{\omega}^{'}(\sin\theta_{\omega}^{'} p_x + \cos\theta_{\omega}^{'} p_z). \tag{A.43}$$

If we now include Eqs. A.42 and A.43 in Eqs. A.28–A.31, they become

$$R_x E_R = f_x E_f^{'} + b_x Q^{'}, \tag{A.44}$$

$$R_y k_{Rz} E_R = f_y k_{fz} E_f^{'}, \text{ since } b_y = 0 \quad \text{(A.45)}$$

$$R_y E_R = f_y E_f^{'}, \text{ since } p_y = 0, \quad \text{(A.46)}$$

$$(R_x k_{Rz} - R_z k_{Rx}) E_R = (f_x k_{fz} - f_z k_{fx}) E_f^{'} + (p_x k_{bz} - p_z k_{bx}) Q^{'}. \quad \text{(A.47)}$$

It follows from Eqs. A.45 and A.46 that $R_y = f_y = 0$. Given that $R_x = \cos(-\theta)$ and $f_x = \cos(\theta_{2\omega}^{'})$, Eq. A.44 becomes

$$-E_R \cos\theta = \cos\theta_{2\omega}^{'} E_f^{'} + b_x Q^{'}. \quad \text{(A.48)}$$

As for Eq. A.47, the left-hand side can be written as

$$(R_x k_{Rz} - R_z k_{Rx}) E_R = |k_R| (\cos\theta \cos\theta + \sin\theta \sin\theta) E_R = (2\omega/c) E_R.$$

Concerning the first term (free harmonic) of the right-hand side,

$$(f_x k_{fz} - f_z k_{fx}) E_f^{'} = (2\omega/c) n^{2\omega} E_f^{'}.$$

And, for the bound harmonic wave term,

$$(p_x k_{bz} - p_z k_{bx}) Q^{'} = (2\omega/c) n^{\omega} (p_x \cos\theta_{\omega}^{'} - p_z \sin\theta_{\omega}^{'}).$$

So, by matching and simplifying terms it is deduced that

$$E_R = n^{2\omega} E_f^{'} + n^{\omega} Q^{'} (p_x \cos\theta_{\omega}^{'} - p_z \sin\theta_{\omega}^{'}). \quad \text{(A.49)}$$

By removing E_R between Equations A.48 and A.49, and isolating the amplitude of the free harmonic wave inside the sample, we obtain

$$E_f^{'} = -Q^{'} \frac{b_x + n^{\omega} \cos\theta (p_x \cos\theta_{\omega}^{'} - p_z \sin\theta_{\omega}^{'})}{\cos\theta_{2\omega}^{'} + n^{2\omega} \cos\theta} = -\beta Q^{'} \quad \text{(A.50)}$$

Again, by substituting the expression for the harmonic field in the material, Eq. A.12, we have

$$\boxed{\vec{E}_{2\omega}^{'} = -\hat{e}_f \beta Q^{'} \exp\left(i\vec{k}_f \cdot \vec{r}\right) + \hat{e}_b Q^{'} \exp\left(i\vec{k}_b \cdot \vec{r}\right)} \quad \text{(A.51)}$$

Boundary Conditions at the Output Surface of the Sample ($Z = L$)

The harmonic waves are transmitted through this surface to the air, so that there is a free wave $E_{2\omega}^{''}$ with the wave vector $k_{2\omega}^{''}$:

$$\vec{\mathrm{k}}_{2\omega}^{''} = 2\vec{\mathrm{k}}_{\omega}$$

We assume that there has not been any reflection for the bound wave. The free reflected wave at 2ω will be $E_f^{'R}$ and its wave vector $k_f^{'R}$, where

$$\theta_f^{'R} = -\theta_{2\omega}^{'}, \quad \left|k_f^{'R}\right| = n_{2\omega}^{'}(2\omega/c).$$

As we did in the $Z = 0$ limit, we will consider, on the one hand, when the nonlinear polarization is perpendicular to the plane of incidence, XZ, and on the other, when it is parallel.

Nonlinear Polarization Perpendicular to the Plane of Incidence (XZ): $p_y = 1$; $b_y = 1$; $f_y = 1$

By imposing the continuity condition on the tangential components of the electric and magnetic fields at the output of the sample ($r = L$) we have

$$E'_f \exp(i\vec{k}_f \hat{Z}L) + Q' \exp(i\vec{k}_b \hat{Z}L) + E'^R_f \exp(-i\vec{k}_f \hat{Z}L) = E''_{2\omega}(2i\vec{k}_\omega \hat{Z}L) \quad \text{(A.52)}$$

$$\begin{aligned} &-n^{2\omega}\cos\theta'_{2\omega}E'_f \exp(i\vec{k}_f \hat{Z}L) - n^{\omega}\cos\theta'_{\omega}Q' \exp(i\vec{k}_b \hat{Z}L) \\ &+n^{2\omega}\cos\theta'_{2\omega}E'^R_f \exp(i\vec{k}_f \hat{Z}L) = -\cos\theta\, E''_{2\omega}\exp(2i\vec{k}_\omega \hat{Z}L) \end{aligned} \quad \text{(A.53)}$$

Removing E'^R_f between the two expressions and working out $E'_{2\omega}$, we obtain

$$\begin{aligned} E''_{2\omega}\exp(i2\vec{k}_\omega \hat{Z}L) = &\left(\frac{2n^{2\omega}\cos\theta'_{2\omega}}{\cos\theta + n^{2\omega}\cos\theta'_{2\omega}}\right)E'_f \exp(i\vec{k}_f \hat{Z}L) \\ &+\left(\frac{n^{\omega}\cos\theta'_{\omega} + n^{2\omega}\cos\theta'_{2\omega}}{\cos\theta + n^{2\omega}\cos\theta'_{2\omega}}\right)Q' \exp(i\vec{k}_b \hat{Z}L) \end{aligned} \quad \text{(A.54)}$$

Having calculated the electric field corresponding to the second harmonic wave generated in the material, the intensity magnitude directly related to what is measured in a SHG experiment can be calculated. For this, we calculate the time average of the Poynting vector, which is the energy flow per area and time unit, that is, the intensity

$$\vec{S} = \left(\frac{1}{2}\right)\text{Re}\left(\vec{E}''_{2\omega} \times \vec{H}''^{*}_{2\omega}\right) \text{ and } \left\langle \vec{S} \right\rangle = I''_{2\omega} \quad \text{(A.55)}$$

$$I''_{2\omega} = \left(\frac{c}{8\pi}\right)\left|E''_{2\omega}\right|^2 \text{ in c.g.s.units.} \quad \text{(A.56)}$$

In order to calculate the magnitude squared of the field, each side of Eq. A.54 is multiplied by the corresponding complex conjugate. Furthermore, it should be noted that

$$\vec{k}_f \cdot \hat{Z} = \left|\vec{k}_f\right|\cos\theta'_{2\omega} = n^{2\omega}\frac{4\pi}{\lambda}\cos\theta'_{2\omega} \text{ and } \vec{k}_b \cdot \hat{Z} = n^{\omega}\frac{4\pi}{\lambda}\cos\theta'_{\omega}.$$

The factors before the exponentials on the right-hand side of Eq. A.54, will be named A and B to simplify the expressions now in the following way:

$$\begin{aligned} |\mathrm{E}_{2\omega'}|^2 &= [Ae^{i4\pi Ln^{2\omega}\cos\theta'_{2\omega}/\lambda} + Be^{-i4\pi Ln^{\omega}\cos\theta'_{\omega}/\lambda}] \\ &\quad \times [Ae^{-i4\pi Ln^{2\omega}\cos\theta'_{2\omega}/\lambda} + Be^{i4\pi Ln^{\omega}\cos\theta'_{\omega}/\lambda}] \\ &= A^2 + B^2 + AB[e^{i4\pi L(n^{2\omega}\cos\theta'_{2\omega} - n^{\omega}\cos\theta'_{\omega})/\lambda} \\ &\quad + e^{-i4\pi L(n^{2\omega}\cos\theta'_{2\omega} - n^{\omega}\cos\theta'_{\omega})/\lambda}] \end{aligned} \quad \text{(A.57)}$$

The term in brackets may be simplified ($2\cos\theta = e^{i\theta} + e^{-i\theta}$), so there remains an oscillating term corresponding to

$$2\cos\left(n^{\omega}\cos\theta'_{\omega} - n^{2\omega}\cos\theta'_{2\omega}\right)\frac{4\pi}{\lambda}L = \sin^2[\left(n^{\omega}\cos\theta'_{\omega} - n^{2\omega}\cos\theta'_{2\omega}\right)2\pi L/\lambda]$$
$$= \sin^2\Psi.$$

The ratio $l_c(\theta) = \frac{\lambda}{4(n^{\omega}\cos\theta'_{\omega} - n^{2\omega}\cos\theta'_{2\omega})}$ is identified as the coherence length (depending on the angle of incidence) for this SHG process. The argument of the previous oscillating function can then be written as $\Psi = \frac{\pi L}{2l_c(\theta)}$.

By substituting into the $|E_{2\omega}'|^2$ expression the values of E_f' and Q' 'given by Eqs. A.39 and A.14, it is straightforward to check that the sum of the non-oscillating terms is several orders of magnitude lower than the amplitude of the oscillating term, so it can be neglected. Thus, the intensity of the second harmonic is given by

$$\boxed{I''_{2\omega} = \left(\frac{c}{8\pi}\right)\left(\frac{16\pi^2\left|P'^2_{2\omega}\right|}{\left((n^{\omega})^2 - (n^{2\omega})^2\right)^2}\right)\left(8n^{2\omega}\cos\theta'_{2\omega}\right) \times \frac{\left(\cos\theta + n^{\omega}\cos\theta'_{\omega}\right)\left(n^{\omega}\cos\theta'_{\omega} + n^{2\omega}\cos\theta'_{2\omega}\right)}{\left(\cos\theta + n^{2\omega}\cos\theta'_{2\omega}\right)^3}\sin^2\psi} \quad \text{(A.58)}$$

Nonlinear Polarization Parallel to the Plane of Incidence (XZ): $p_z = 0$; $b_z = 0$

By following the reasoning that we have used during this development, if we apply the continuity conditions to the electric and magnetic fields we have

$$\cos\theta E''_{2\omega}(2i\vec{k}_{\omega}\hat{Z}L) = \cos\theta'_{2\omega}E'_f\exp(i\vec{k}_f\hat{Z}L) + b_x Q'\exp(i\vec{k}_b\hat{Z}L) - E'^R_f\cos\theta'_{2\omega}\exp(-i\vec{k}_f\hat{Z}L) \quad \text{(A.59)}$$

$$E''_{2\omega}(2i\vec{k}_{\omega}\hat{Z}L) = n^{2\omega}E'_f\exp(i\vec{k}_f\hat{Z}L) + n^{\omega}Q'(p_x\cos\theta'_{\omega} - p_z\sin\theta'_{\omega})\exp(i\vec{k}_b\hat{Z}L) + n^{2\omega}E'^R_f\cos\theta'_{2\omega}\exp(-i\vec{k}_f\hat{Z}L) \quad \text{(A.60)}$$

Removing $E_f'^R$ from these equations, it is deduced that

$$E''_{2\omega}\exp(2i\vec{k}_{\omega}\hat{Z}L) = \left(\frac{2n^{2\omega}\cos\theta'_{2\omega}}{n^{2\omega}\cos\theta + \cos\theta'_{2\omega}}\right)E'_f\exp(i\vec{k}_f\hat{Z}L) + \left(\frac{n^{2\omega}b_x + n^{\omega}\cos\theta'_{2\omega}(\cos\theta'_{\omega}p_x - \sin\theta'_{\omega}p_z)}{\cos\theta'_{2\omega} + n^{2\omega}\cos\theta}\right)Q'\exp(i\vec{k}_b\hat{Z}L) \quad \text{(A.61)}$$

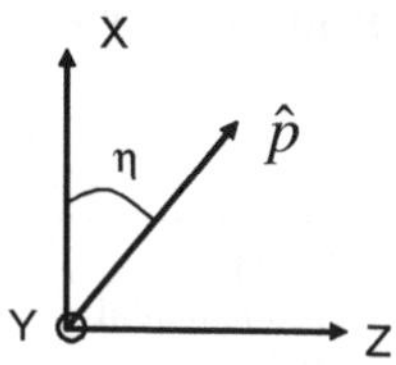

Fig. A.2 Definition of angle η

By using again Eq. A.56 and substituting for the amplitudes of the free and bound waves now, given by Eqs. A.14 and A.50, we reach the expression for the second harmonic intensity at the exit of the film. As in the case just described, only the oscillating term remains:

$$I''_{2\omega} = \left(\frac{c}{8\pi}\right)\left(\frac{16\pi^2\left|P'_{2\omega}\right|^2}{\left((n^{\omega})^2-(n^{2\omega})^2\right)^2}\right)(8n^{2\omega}\cos\theta'_{2\omega})$$
$$\frac{\left[b_x + n^{\omega}\cos\theta'_{\omega}(p_x\cos\theta'_{\omega} - p_z\sin\theta'_{\omega})\right]\left[n^{2\omega}b_x + n^{\omega}\cos\theta'_{2\omega}(p_x\cos\theta'_{\omega} - p_z\sin\theta'_{\omega})\right]}{\left(n^{2\omega}\cos\theta + \cos\theta'_{2\omega}\right)^3}\sin^2\psi \tag{A.62}$$

This equation can be simplified if we recall the expression for the X component of the unit vector $\hat{e}_b$ (b_x, Eq. A.42), and we consider that there is little dispersion, i.e. $n^{\omega} \approx n^{2\omega}$. Under these conditions, $b_x = p_x - \sin\theta'_{\omega}(\sin\theta'_{\omega}p_x + \cos\theta'_{\omega}p_z)$, which is equal to $b_x = \cos\theta'_{\omega}(p_x\cos\theta'_{\omega} - p_z\sin\theta'_{\omega})$.

We now define an angle called η between the polarization vector and the X direction (Fig. A.2).

The unit vector components will be

$$p_x = \cos\eta \text{ and } p_z = \sin\eta,$$

so

$$(p_x\cos\theta'_{\omega} - p_z\sin\theta'_{\omega}) = (\cos\eta\cos\theta'_{\omega} - \sin\eta\sin\theta'_{\omega}) = \cos(\theta'_{\omega} - \eta).$$

Introducing the above in Eq. A.62 allows the expression to be reduced to

$$I''_{2\omega} = \left(\frac{c}{8\pi}\right)\left(\frac{16\pi^2\left|P'_{2\omega}\right|^2}{\left((n^{\omega})^2-(n^{2\omega})^2\right)^2}\right)8n^{2\omega}\cos\theta'_{2\omega}\cos^2(\theta'_{\omega} - \eta)$$
$$\times\frac{(n^{\omega}\cos\theta + \cos\theta'_{\omega})(n^{2\omega}\cos\theta'_{\omega} + \cos\theta'_{2\omega})}{(n^{2\omega}\cos\theta + \cos\theta'_{2\omega})^3}\sin^2\psi \tag{A.63}$$

In view of Eqs. A.58 and A.63, a general expression for the second harmonic intensity at the exit of the film can be written, which relates it to the fundamental electric field amplitude:

$$I''_{2\omega} = \left(\frac{8\pi\ c}{\left((n^{\omega})^2-(n^{2\omega})^2\right)^2}\right)d^2p^2(\theta)\left|E'_{\omega}\right|^4 t''^4_{\omega}T''_{2\omega}\sin^2\psi \tag{A.64}$$

Table A.1 Transmission factors

	Fundamental, parallel to the plane of incidence P^{ω}	Perpendicular to the plane of incidence S^{ω}
Air/sample, fundamental light	$t'_{\omega} = \frac{2\cos\theta}{\left(n^{\omega}\cos\theta + \cos\theta'_{\omega}\right)}$	$t'_{\omega} = \frac{2\cos\theta}{\left(n^{\omega}\cos\theta'_{\omega} + \cos\theta\right)}$
	Harmonic, parallel to the plane of incidence $P^{2\omega}$	Perpendicular to the plane of incidence $S^{2\omega}$
Sample/air, harmonic light	$T'_{2\omega} = 2n^{2\omega}\cos\theta'_{2\omega} \frac{(n^{\omega}\cos\theta + \cos\theta'_{\omega})(n^{2\omega}\cos\theta'_{\omega} + \cos\theta'_{2\omega})}{(n^{2\omega}\cos\theta + \cos\theta'_{2\omega})^3}$	$T'_{2\omega} = 2n^{2\omega}\cos\theta'_{2\omega} \frac{(\cos\theta + n^{\omega}\cos\theta'_{\omega})(n^{\omega}\cos\theta'_{\omega} + n^{2\omega}\cos\theta'_{2\omega})}{(\cos\theta + n^{2\omega}\cos\theta'_{2\omega})^3}$

In this expression, d is the nonlinear coefficient involved in the measurement and $p(\theta)$ is a projection factor, which depends on the nonlinear tensor form and the polarization direction $P_{2\omega}'$, such that $p(\theta) = p_1 p_2$, where p_1 is defined by the expression for the polarization amplitude as a function of the fundamental field amplitude: $\left|P'_{2\omega}\right| = p_1 d \left|E'_{\omega}\right|^2$, and p_2 depends on the direction of the nonlinear polarization:

$p_2 = 1$ if it is perpendicular to the plane of incidence and
$p_2 = cos(\theta_{\omega}' - \eta)$ if it is parallel to the plane of incidence

Furthermore, as mentioned previously, we consider the transmission factors air/sample of the fundamental wave $\left|E_{\omega'}\right| = t_{\omega'} \left|E_{\omega}\right|$ and sample/air factor for the harmonic wave $T_{2\omega}'$. See Table A.1.

In view of Eq. A.64, it follows that the model proposed by Jerphagnon and Kurtz predicts that the second harmonic intensity is an oscillating function with the light path in the material, that is, with angle θ, varying when the sample is rotated. The periodic variation of the signal is modulated by the presence of the projection factor $p(\theta)$, mainly responsible for the "envelope" shape of the Maker fringes in the SHG experiments in solids.

Correction for Highly Birefringent Samples

As mentioned, in this thesis the Herman and Hayden method (hereafter HH) has been used to fit the data corresponding to polymer films owing to their remarkable birefringence. The measurement configurations were $P^{\omega}P^{2\omega}$ and $S^{\omega}P^{2\omega}$. The first correction that we will consider is the inclusion of the mentioned *walk-off* angle. The formula to calculate it is [3]

$$\tan\gamma_{\omega} = \frac{(n^{\omega})^2}{2}\left[\frac{1}{(n_o^{\omega})^2} - \frac{1}{(n_e^{\omega})^2}\right]\sin 2\theta'_{\omega}, \quad \text{For the bound harmonic wave.}$$

$$\tan\gamma_{2\omega} = \frac{(n^{2\omega})^2}{2}\left[\frac{1}{(n_o^{2\omega})^2} - \frac{1}{(n_e^{2\omega})^2}\right]\sin 2\theta'_{2\omega} \quad \text{For the free harmonic wave.} \tag{A.65}$$

In the present case, we have to define a new refractive index for the bound harmonic wave that will be dependent on the angle of refraction θ'_ω :

$$n^{2\omega l}(\theta'_\omega) = \left(\frac{\cos^2 \theta'_\omega}{n_o^{2\omega}} + \frac{\sin^2 \theta'_\omega}{n_e^{2\omega}} \right)^{-1/2}. \tag{A.66}$$

The HH model uses the same air/sample transmission factors for the fundamental beam as the JK model. However, it includes reflection losses at the sample/glass interface for the harmonic, so it introduces two new factors, with an additional correctionfor the *walk-off* angle.

The incident light polarization is perpendicular, S^ω

The air/sample transmission factor, $t_\omega^{a\to m}$, will be the same as in the JK model, because $\gamma_\omega = 0$, as the bound harmonic wave is polarized perpendicular to the optical axis and the index it will "see" will be n_o^{ω}.The transmission factor sample/substrate will be

$$T_{2\omega}^{m\to s} = \frac{2n^{2\omega} \cos\gamma_{2\omega} \cos(\theta'_{2\omega} - \gamma_{2\omega})}{n_S^{2\omega} \cos(\theta'_{2\omega} - \gamma_{2\omega}) + n^{2\omega} \cos\theta'_{S2\omega} \cos\gamma_{2\omega}} \tag{A.67}$$

$n_S^{2\omega}$ substrate refractive index,
$\theta_{S2\omega'}$ propagation angle in the substrate.

$$\sin\theta_{S2\omega} = \frac{n^{2\omega} \sin\theta'_{2\omega}}{n_S^{2\omega}} \tag{A.68}$$

$$T_{2\omega}^{s\to a} = \frac{2n_S^{2\omega} \cos\theta'_{S2\omega}}{n_S^{2\omega} \cos\theta + \cos\theta'_{S2\omega}} \tag{A.69}$$

The incident light polarization is parallel, P_ω

The *walk-off* angle has to be taken into account in oriented films, since the fundamental wave propagates such that the direction of its polarization vector forms an angle (variable during the measurement) with the optical axis. Then the transmission factor is no longer the same:

$$t_\omega^{a\to m} = \frac{2\cos\theta}{\cos\left(\theta'_\omega - \gamma_\omega\right) + n^\omega \cos\theta \cos\gamma_\omega} \tag{A.70}$$

The $T'_{2\omega}$ factors have the same form as in the JK model.

Moreover, an effective coefficient, d_{eff}, is defined, whose relationship with the parameters of the expression in the JK model is $d_{eff} = p(\theta)\, d$.

Finally, by introducing these modifications, the expression for the second harmonic power at the exit of the birefringent material is obtained:

$$\boxed{\begin{aligned} P_{2\omega} &= \frac{128\pi^3}{cA} d_{eff}^2 \left(\frac{n^{2\omega l}}{n^{2\omega}}\right)^4 \left(\frac{(n^\omega)^2 - \left(n^{2\omega}\right)^2}{(n^\omega)^2 - \left(n^{2\omega l}\right)^2}\right) \sin^2\psi \\ &\quad \frac{\left[t_\omega^{a\to m}\right]^4 \left[T_{2\omega}^{m\to s}\right]^2 \left[T_{2\omega}^{s\to a}\right]^2}{\cos^2\gamma_{2\omega} \cos^2(\theta'_{2\omega} - \gamma_{2\omega})(n^\omega \cos\theta'_\omega - n^{2\omega} \cos\theta'_{2\omega})^2} P_\omega^2 \end{aligned}} \tag{A.71}$$

Appendix B
Expression for the Second Harmonic Intensity at the Exit of the EFISH Cell

The second harmonic intensity at the exit of the measurement cell for liquids is an oscillating function of the distance travelled by the light, as it was in the case of polymeric films. The derivation of the second harmonic intensity expression at the output of the cell follows Oudar's work [4]. As we saw, in that case the wave propagation equation for frequency 2ω is

$$\nabla \times \nabla \times \vec{E_{2\omega}} + \frac{\varepsilon_{2\omega}}{c^2}\frac{\partial^2 \vec{E_{2\omega}}}{\partial t^2} = \frac{-4\pi}{c^2}\frac{\partial^2 \vec{P_{2\omega}^{NL}}}{\partial t^2}. \tag{B.1}$$

In the case of EFISH, all the electric fields are parallel to the laboratory Z-axis (Fig. B.1), since the static electric field is applied in that direction, and the fundamental wave reaches the solution with a polarization perpendicular to the plane of incidence, X being the propagation direction.
So, the expression for the polarization as a function of the fields is

$$P_{2\omega}^{NL} = \Gamma_{ZZZZ}(-2\omega, \omega, \omega, 0)E_Z^{\omega}E_Z^{\omega}E_Z^{0} = dE_Z^{\omega}E_Z^{\omega} \text{ with } d \equiv \Gamma_{ZZZZ}E_0. \tag{B.2}$$

We recall here the expressions for the vectors of the fundamental wave and harmonics defined in Appendix A:

$$\left|\vec{k_\omega}\right| = n^{\omega}\left(\frac{\omega}{c}\right), \quad \left|\vec{k_f}\right| = n^{2\omega}\left(\frac{2\omega}{c}\right), \quad \left|\vec{k_b}\right| = n^{\omega}\left(\frac{2\omega}{c}\right).$$

A solution for the wave equation *B. 1* would be

$$\vec{E_{2\omega}}(x,t) = E^{2\omega}(x)e^{-i(2\omega)t}\,\vec{Z} \tag{B.3}$$

and for the fundamental wave

$$\vec{E_\omega}(x,t) = E^{\omega}(x)e^{-i\omega t}\,\vec{Z} = E^{\omega}e^{-i\omega(t-\frac{n^{\omega}}{c}x)}\,\vec{Z} \tag{B.4}$$

R. Alicante, *Photoinduced Modifications of the Nonlinear Optical Response in Liquid Crystalline Azopolymers*, Springer Theses, DOI: 10.1007/978-3-642-31756-9,

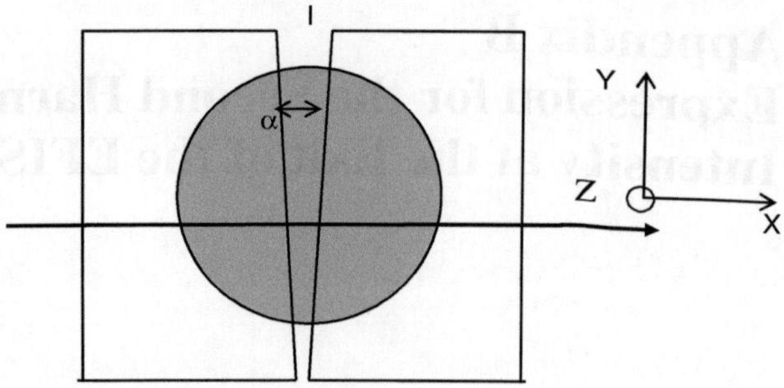

Fig. B.1 Detail of the EFISH cell and the laboratory axes

since $E^{\omega}(x) = E^{\omega} e^{ik_{\omega}x}$.
With the above, Eq. B.1 becomes

$$\frac{\partial^2 E^{2\omega}}{\partial x^2} + k_f^2 E^{2\omega} = \frac{-16\pi\omega^2}{c^2} d(E^{\omega})^2 e^{ik_b x}, \tag{B.5}$$

whose solution is presented as a sum of two terms in an analogous way to that seen in Appendix A:

$$E^{2\omega}(x) = E_f e^{ik_f x} + E_b e^{ik_b x}.$$

Under the usual conditions of an EFISH experiment, the amplitude of the bound wave E_b, is proportional to the nonlinear polarization:

$$E^{2\omega}(x) = E_f e^{ik_f x} - \frac{4\pi}{(n_{2\omega})^2 - (n_{\omega})^2} d(E^{\omega})^2 e^{ik_b x}. \tag{B.6}$$

This field is transmitted through the different media presented in the figure. At the boundary between two media the reflected and transmitted waves will appear, whose amplitudes are calculated by the boundary conditions of the fields $\vec{E}_{2\omega}$ and $\vec{H}_{2\omega}$, in a similar way to the case of the films described in Appendix A, that is, by matching tangential components at the interface between the media. The separation surfaces that the wave will find will be air/quartz, quartz/liquid, liquid/quartz and quartz/air. At the second's window output, the amplitude of the transmitted harmonic field can be expressed in terms of the bound wave amplitudes in the quartz, E_b^Q and in liquid E_b^L, as well as of the corresponding transmission factors:

$$E^{2\omega} = t_{2\omega}^{q\to a}(T_Q E_b^Q - T_L E_b^L)(e^{ikfl} - e^{ikbl}) \tag{B.7}$$

$$E_b^Q = \frac{-4\pi}{\left(n_q^{2\omega}\right)^2 - \left(n_q^{\omega}\right)^2} \Gamma_G E_0 \left(E_{\omega} t_{2\omega}^{q\to a}\right)^2 \tag{B.8}$$

$$E_b^L = \frac{-4\pi}{\left(n_l^{2\omega}\right)^2 - \left(n_l^{\omega}\right)^2} \Gamma_L E_0 \left(E_{\omega} t_{2\omega}^{a\to q} t_{2\omega}^{q\to l}\right)^2 \tag{B.9}$$

Where t and T factors are defined as in Table B.1.

Table B.1 Transmission factors

Quartz/Air	$t_{2\omega}^{q\to a} = \frac{2n_q^{2\omega}}{1+n_q^{2\omega}}$	Quartz/Liquid	$T_L = \frac{n_l^{\omega}+n_l^{2\omega}}{n_q^{2\omega}+n_l^{2\omega}}$
Air/Quartz	$t_{2\omega}^{a\to q} = \frac{2}{1+n_q^{\omega}}$	Liquid/Quartz	$T_Q = \frac{n_q^{\omega}+n_l^{2\omega}}{n_q^{2\omega}+n_l^{2\omega}}$

From the expression for the harmonic field in Eq. B.7, the intensity is calculated using Eq. A.56:

$$I_{2\omega} = \frac{c}{8\pi}\left|E_{2\omega}\right|^2 = 2I_M^2 \sin^2\left(\frac{\Delta\phi}{2}\right), \tag{B.10}$$

where

$$I_{2\omega} = 2I_M^L \sin^2\left(\frac{\Delta\phi}{2}\right), \tag{B.11}$$

$$I_M^L = \frac{c}{4\pi}\left[t_{2\omega}^q(T_Q E_b^q - T_L E_b^L)\right]^2, \tag{B.12}$$

$$\Delta\phi = l(k_b - k_f). \tag{B.13}$$

The coherence length is defined as

$$l_c = \frac{\pi}{k_{2\omega} - 2k_\omega} = \frac{\lambda_v}{4(n^{2\omega} - n^\omega)} \tag{B.14}$$

where λ_v is the wavelength in the vacuum.

Thus, finally, we have that

$$I_{2\omega} = 2I_M^L \sin^2\left(\frac{\pi l}{2l_c}\right), \tag{B.15}$$

which is the expression used in fitting the Maker fringes to obtain I_M^L and l_c, as explained in Chap. 2 of this work.

I_M^L can be written so that the value of the nonlinear susceptibility of the liquid appears explicitly. Thus, replacing E_b^L, E_b^Q, T_L and T_Q in Eq. B.12, we find

$$\sqrt{I_M^L} = C\left[\Gamma_L l_c - K\right]E_0, \ where\, C\, and\, K\, are \tag{B.16}$$

$$C \equiv \sqrt{c/4\pi}\,\frac{-64\pi}{\lambda_v}\,t_{2\omega}^{q\to a}\,\frac{(n_q^\omega)^2}{(n_q^\omega+n_l^\omega)^2}\,\frac{(E^\omega t_\omega)^2}{(n_q^{2\omega}+n_l^{2\omega})^2},$$

$$K \equiv \frac{\lambda_v \Gamma_q}{16}\,\frac{(n_q^\omega + n_l^{2\omega})}{(n_q^\omega)^2}\,\frac{(n_q^\omega + n_l^\omega)^2}{\left[(n_q^{2\omega})^2 - (n_q^\omega)^2\right]}.$$

Table B.2 Values of the parameters needed for the experimental calculation of $\mu\beta$ in different solvents

	Quartz windows (Suprasil)		Dichloromethane (CH_2Cl_2)		Chloroform ($CHCl_3$)		Dimethylsulfoxide DMSO ($(CH_3)_2SO$)	
λ[nm]	1064	1906	1064	1906	1064	1906	1064	1906
n^{ω}	1.450	1.440	1.420	1.407	1.438	1.435	Not used	1.462
$n^{2\omega}$	1.460	1.450	1.430	1.410	1.450	1.439		1.467
Γ [10^{-14} esu]	3.3	2.9	11.7	8	8.7	6.4	Not used	−5.20
ε	–	–	9,080		4.806		46.7	
ρ [g/cm^3]	–	–	1.322		1.483		1.095	
K [10^{-14} esu]	–	–	75.6	115	77.6	137.0	Not used	122
l_c [μm]	–	–	27	136	22.5	120.8	Not used	95

Table B.2 summarizes the parameters needed to calculate $\mu\beta$ of the molecules in the different solvents and for the wavelengths used in this work.

Appendix C
Order Parameters

In the context of this work, the order parameters characterize the degree of order of a chromophore distribution in a polymer matrix. In a system composed of uniaxial molecules with cylindrical symmetry, the order parameters correspond to the coefficients of the orientational distribution function $F(\theta, \varphi)$ as a series of Legendre polynomials ($P_n(\cos\,\theta)$), where θ the angle between the axis and the macroscopic molecular axis, the only significant parameter (see Fig. C.1).

$$F(\theta) = 2\pi F(\theta, \varphi) = \sum_{i=0}^{\infty} \left(\frac{2i+1}{2}\right) < P_i > P_{i(\cos\theta)},$$

$$P_0 = 1 \quad P_1 = \cos\theta \quad P_2 = \frac{3}{2}\cos^2\theta - \frac{1}{2} \quad P_3 = \frac{5}{2}\cos^3\theta - \frac{3}{2}\cos\theta.;\ P_4, P_5 \ldots$$

A centrosymmetric system involves only even-order terms. On the other hand, if there is polar order it is necessary to consider the coefficients of both the odd- and even-order terms.

The angular brackets < > refer to the orientational average of the corresponding magnitude. To calculate this, we note that the probability of finding a molecule oriented in a certain interval dθ around θ is given by $F(\theta)$ sin θ dθ. So $<P_1>$ is calculated as

$$P_1 = \langle\cos\theta\rangle = \frac{\int_0^\pi \cos\theta \; F(\theta)\sin\theta\, d\theta}{\int_0^\pi F(\theta)\sin\theta\, d\theta}$$

Throughout this report two types of arrangements have been mentioned, an axial orientation determined by the "long" axis of the molecules around a preferred direction (which corresponds to the parameter $<P_2>$) and another polar one, in which the orientation of the dipole moment of the molecules with respect to a reference direction is decisive, where $<P_1>$ and $<P_3>$ also have to be taken into account. Below, we will see how to obtain the order parameters from experimental data.

R. Alicante, *Photoinduced Modifications of the Nonlinear Optical Response in Liquid Crystalline Azopolymers*, Springer Theses, DOI: 10.1007/978-3-642-31756-9,

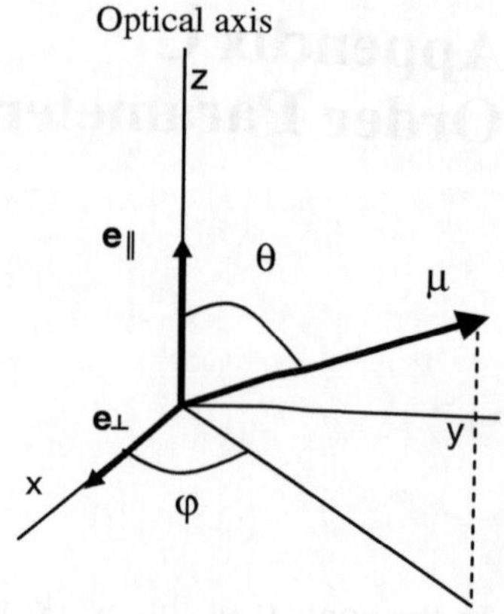

Fig. C.1 Definition of the angles between the molecule dipole moment and the coordinate axes

Axial Order Parameter

The axial order parameter is defined as the statistical average of the Legendre polynomial of second order:

$P_2 = <P_2(\cos\theta)> = \frac{1}{2} <3\cos^2\theta - 1>$, so that $<\cos^2\theta> = \frac{1+2P_2}{3}$.

This parameter has been calculated by two experimental methods in the work reported in this thesis. One of them is based on UV-vis absorption measurements and the other on refractive index measurements of the films. It is obvious that neither of these techniques distinguish between dipoles pointing in the same direction or with opposite orientations.

From Absorption Measurements

The probability of absorption of a photon polarized along the $\hat{\boldsymbol{e}}$ vector by a dipole μ is proportional to $(\boldsymbol{\mu}.\hat{\boldsymbol{e}})^2$ (see Fig. C.1). Therefore, for a uniaxial distribution of dipoles, the absorption of polarized light in a direction parallel and perpendicular to the optical axis will be [5]

$$A_{\parallel} = \kappa\left\langle(\vec{\mu}\cdot\hat{e}_{\parallel})^2\right\rangle = \kappa\left\langle\cos^2\theta\right\rangle = \kappa\frac{1+2P_2}{3} \tag{C.1}$$

$$\begin{aligned} A_{\perp} &= \kappa\left\langle(\vec{\mu}\cdot\hat{e}_{\perp})^2\right\rangle = \kappa\left\langle\sin^2\theta\cos^2\varphi\right\rangle \\ &= \kappa\int\limits_{\theta=0}^{\pi}\int\limits_{\varphi=0}^{2\pi}\sin^2\theta\cos^2\varphi\frac{F(\theta)}{2\pi}\sin\theta\,d\varphi\,d\theta \\ &= \kappa\frac{1}{2}\int\limits_{\theta=0}^{\pi}\sin^2\theta F(\theta)\,\sin\theta\,d\theta = \kappa\frac{1}{2}\langle 1-\cos^2\theta\rangle = \kappa\frac{1-P_2}{3} \end{aligned} \tag{C.2}$$

where κ is a constant.

By using the two expressions to remove the constant, we obtain the expression for the axial order parameter P_2 as a function of the polarized absorption [6]:

$$P_2 = \frac{A_{\parallel} - A_{\perp}}{A_{\parallel} + 2A_{\perp}}. \tag{C.3}$$

If we define A_0 as the initial absorbance of the sample (isotropic) measured at normal incidence and $A_{\perp}$ as the absorbance of the sample axially oriented, measured with light polarized perpendicularly to the orientation direction (normal incidence), it must be satisfied, taking into account the three directions of the film, that

$$A_{\parallel} + 2A_{\perp} = 3A_0, \tag{C.4}$$

and this expression can be reduced to

$$P_2 = 1 - \frac{A_{\perp}}{A_0} \tag{C.5}$$

From Refractive Index Measurements

The calculation of the order parameter P_2 from refractive indices was performed by following the method described by Page et al. [7]. This method is based on the fact that the contribution to the refractive index of the charge transfer band is proportional to the absorbance. The refractive index has therefore a frequency-independent contribution, n_0, and another one with resonant character, associated with the charge transfer transition at frequency ω_0. We then have that

$$n(\omega) = n_0 + \delta n(\omega) \text{ and } \delta n(\omega) \propto \frac{Const}{\omega_0^2 - \omega^2}. \tag{C.6}$$

The frequency ω_0, at which the absorption is maximal, is obtained from the UV–vis absorption spectrum of the film.

When the chromophores are oriented axially in the film, the components parallel and perpendicular to the axis, $\delta n_{\parallel}$ and $\delta n_{\perp}$, respectively, vary with respect to their value in the isotropic system δn_0, but the frequency dependence is the same for all the components, so that

$$\delta n_{\parallel} + 2\delta n_{\perp} = 3\delta n_0. \tag{C.7}$$

According to Eq. C.2, the order parameter can then be expressed as

$$P_2 = \frac{\delta n_{\parallel} - \delta n_{\perp}}{\delta n_{\parallel} + 2\delta n_{\perp}}. \tag{C.8}$$

The technique used to measure the refractive indices utilizes TE polarized light (perpendicular to the plane of incidence) and TM (parallel to the plane of incidence). In the event that the axis is perpendicular to the film plane, as in the cases analyzed, we must have

$$\delta n_{TM} = \delta n_{\parallel}$$
$$\delta n_{TE} = \delta n_{\perp}$$

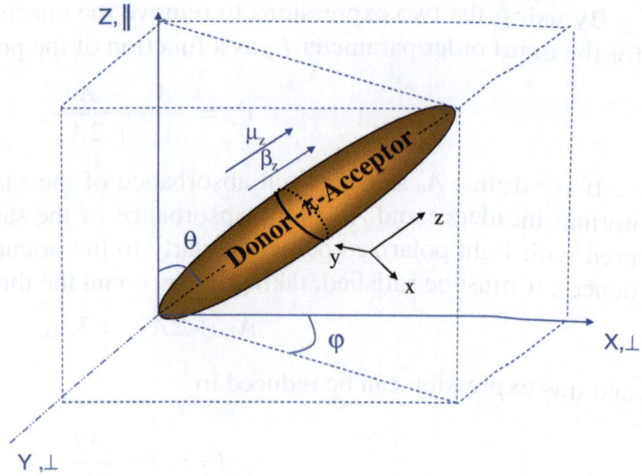

Fig. C.2 Relationship between the molecular dipole moment μ_z, the vector component of the susceptibility β_z, and the molecular coordinates and the macroscopic reference system (Z parallel to the external field and X and Y perpendicular)

Therefore

$$P_2 = \frac{\delta n_{TM} - \delta n_{TE}}{\delta n_{TM} + 2\delta n_{TE}}, \tag{C.9}$$

where $\delta n_{TM} = n_{TM} - n_0$
$\delta n_{TE} = n_{TE} - n_0$

The method used in this thesis is based on the measurement of both n_{TE} and n_{TM} indices at two different wavelengths, (633 and 1306 nm), calculating the average for each wavelength:

$$n(\omega) = \sqrt{\frac{2(n_{TE})^2 + (n_{TM})^2}{3}}. \tag{C.10}$$

Thus, solving the dispersion curve in Eq. C.6 leads to n_0 and the constant. (The dispersion expression has been also used to calculate $n(\omega)$ at the frequencies of interest in the calculation of the nonlinear coefficients corresponding to 1907 and 954 nm.)

Polar Order Parameters

The parameters P_1 and P_3 that describe the polar order may be estimated from the results of SHG measurements. To obtain them, the starting point will be the relationship between macroscopic and microscopic nonlinear coefficients in polymeric films, as discussed in Chap. 1 of this work.

The non-centrosymmetric order of the chromophores is achieved by applying an electric field, which orients the dipoles parallel to it. At the microscopic level it is a very complex problem, which depends on the potential energy of the molecular dipoles, the energy associated with thermal fluctuations of the system, the dielectric properties of the medium, etc. At the macroscopic level, the polar orientation process in an isotropic sample induces $C_{\infty v}$ symmetry, with an axis perpendicular to the plane of the film. The relationship between the axes of the molecule and the film oriented this way are shown in Fig. C.2.

The chromophores are distributed in a conical manner around the *Z*-axis, characterized by the angle θ. When the poling is very weak, the θ distribution will be wide, but each molecule will show a certain tendency to align along the *Z*-axis. As discussed in Chap. 1 of this report, the nonlinear optical response of this kind of system is determined by only two nonlinear coefficients:

$$\chi_{33} = \chi_{zzz} \text{ and } \chi_{31} = \chi_{yyz} = \chi_{xxz} = \chi_{zyy} = \chi_{zxx},$$

whose relationship with the microscopic hyperpolarizability is

$$\chi_{zzz} = Nf <bzzz> = Nf <\cos^3\theta> \beta_z \tag{C.11}$$

For its part, the second coefficient (if the *X*-axis is defined to coincide with the component of the incident light in the plane perpendicular to the *Z*-axis) is

$$\chi_{zxx} = Nf\langle b_{p(ZXX)}\rangle = Nf\langle(\cos\theta)(\sin^2\theta)(\cos^2\varphi)\rangle\beta_z \tag{C.12}$$

In the previous expressions, *f* refers to the corresponding local field factors in each case.

In order to calculate the angular function averages that appear in Eqs. C.11 and C.12, the following distribution function will be used:

$$F(\theta) = e^{-U(\theta)/kT}, \tag{C.13}$$

where $U(\theta)$ is the potential energy of the molecule in the medium in the presence of the electric field. $U(\theta)$ has a contribution due to the potentials associated with local structures or the anisotropy of the medium $U_l(\theta)$, and another contribution due to the interaction of the $\boldsymbol{\mu}$ dipole with the applied field $\boldsymbol{E}$. Therefore

$$U(\theta) = U_1(\theta) - \vec{\mu}\cdot\vec{E} = U_1(\theta) - \mu E\cos\theta. \tag{C.14}$$

Here $E = E_{ext}f(0)$, where E_{ext} is the applied external electric field and $f(0)$ the local field factor at zero frequency.

We shall first determine $P_1 = <\cos\theta>$, which corresponds to the orientation average of a dipole in the system. Using the Boltzmann distribution law and considering an isotropic starting medium ($U_1(\theta) = 0$), we have

$$\langle\cos\theta\rangle = \frac{\int_0^\pi \cos\theta\ F(\theta)\sin\theta\, d\theta}{\int_0^\pi F(\theta)\sin\theta\, d\theta} = \frac{\int_0^\pi \cos\theta\ \exp\left(\frac{\vec{\mu}\vec{E}}{kT}\right)\sin\theta\, d\theta}{\int_0^\pi \exp\left(\frac{\vec{\mu}\vec{E}}{kT}\right)\sin\theta\, d\theta} \tag{C.15}$$

The solution is the first-order Langevin function, $L_1(p)$,

$$\langle\cos\theta\rangle = \coth p - \frac{1}{p} = L_1(p) \tag{C.16}$$

which, expressed in a power series expansion of $p = \mu E/kT$, is

$$L_1(p) = \frac{1}{3}p - \frac{1}{45}p^3 + \frac{2}{945}p^5 - \ldots \tag{C.17}$$

Analogously, we can calculate the statistical average $<\cos^3\theta>$, shown in Eq. C.11, which relates the macroscopic nonlinear coefficient to the molecular one:

$$\langle \cos^3\theta \rangle = \frac{\int_0^\pi \cos^3\theta F(\theta)\ \sin\theta\, d\theta}{\int_0^\pi F(\theta)\sin\theta\, d\theta}, \tag{C.18}$$

whose solution is the third-order Langevin function

$$L_3(p) = \left(1 + \frac{6}{p^2}\right) L_1(p) - \frac{2}{p} \ldots. \tag{C.19}$$

By introducing the Langevin function in Eq. C.11 we have

$$\boxed{\chi^{(2)}_{ZZZ} = N\ f_z^2\,(\omega)\, f_z\,(2\omega)\, \beta_z L_3\,(p)} \tag{C.20}$$

Once the expression for the first nonlinear coefficient ($\chi_{zzz} = \chi_{33}$) is obtained, we can calculate χ_{31} by using Eq. C.12. In the case of a polarly oriented isotropic solid, the plane perpendicular to the polar axis direction should remain isotropic and the term $(\cos^2\varphi)$ should not appear in the Boltzmann average:

$$\langle(\cos\theta)(\sin^2\theta)(\cos^2\varphi)\rangle = \frac{\int_0^\pi F(\theta)\cos\theta\ \sin^3\theta\, d\theta}{\int_0^\pi F(\theta)\sin\theta\, d\theta}\int_0^{2\pi}\cos^2\varphi\, d\varphi = \frac{\langle\cos\theta - \cos^3\theta\rangle}{2}$$

$$= \tfrac{1}{2}(L_1(p) - L_3(p))$$

$$\boxed{\chi^{(2)}_{p(ZXX)} = \frac{N\beta_z f_x^2(\omega) f_z(2\omega)(L_1(p) - L_3(p))}{2}} \tag{C.21}$$

Since the relationships between the Langevin functions $L_n(p)$ and the average values of the Legendre polynomials $<P_n(\cos\theta)>$ are

$$L_1\,(p) \;= < P_1(\cos\theta) >, \tag{C.22}$$

$$L_2\,(p) \;= 1/3\,(2 < P_2\,(\cos\theta) > +1)\,, \tag{C.23}$$

$$L_3\,(p) \;= 1/5\,(2 < P_3\,(\cos\theta) > +3 < P_1\,(\cos\theta) >)\,, \tag{C.24}$$

in the notation used in this thesis we would have

$$\chi_{33} = 2d_{33} = N\beta_z f\ <\cos^3\theta>= N\beta_z f\left(\frac{3}{5}\langle P_1\rangle + \frac{2}{5}\langle P_3\rangle\right), \tag{C.25}$$

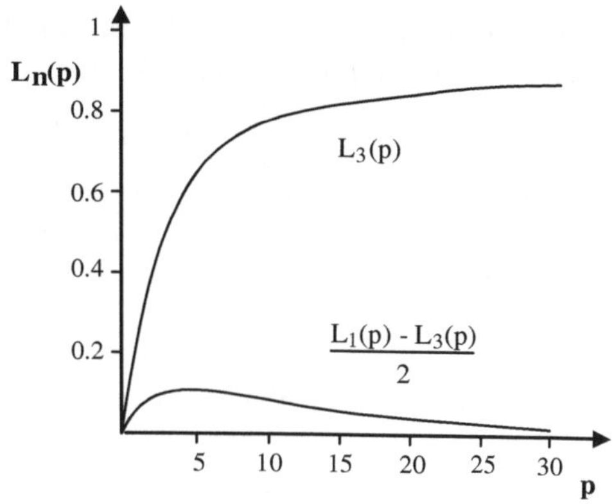

Fig. C.3 The Langevin function $L_3(p)$ and $[L_1(p) - L_3(p)]/2$

$$\chi_{31} = 2d_{31} = N\beta_z f\left(\frac{\langle\cos\theta - \cos^3\theta\rangle}{2}\right) = N\beta_z f\left(\frac{1}{5}\langle P_1\rangle - \frac{1}{5}\langle P_3\rangle\right), \quad \text{(C.26)}$$

and therefore the polar order parameters can be estimated from the nonlinear coefficient values [8]:

$$\langle P_1\rangle = \frac{2(d_{33} + 2d_{31})}{N\beta_z f}, \quad \text{(C.27)}$$

$$\langle P_3\rangle = \frac{2(d_{33} - 3d_{31})}{N\beta_z f}. \quad \text{(C.28)}$$

In the limit of very small fields and low dipole moments, the equations for the nonlinear coefficients take the following form:

$$d_{33} = \frac{Nf\mu E\beta_z}{10kT}, \quad \text{(C.29)}$$

$$d_{31} = \frac{Nf\mu E\beta_z}{30kT}. \quad \text{(C.30)}$$

By comparing the previous two expressions, it can be deduced that for initially isotropic media, where the induced order is due to the thermal poling orientation with low fields, the ratio d_{33}/d_{31} is 3 [9].

In the other limit, for high field values, χ_{33} approaches its saturation value corresponding to $L_3(p) = 1$, and yet χ_{31} approaches zero since it is proportional to $(L_1(p) - L_3(p))/2$, as shown in Fig. C.3.

On the other hand, the anisotropy in the medium, as in the case of liquid crystals, has a significant impact on the degree of poling equilibrium. The principal axis of the liquid crystal is a non-polar axial axis resulting from the alignment due to van der Waals interactions between mesogens, while the external electric field orients these molecules polarly. Therefore, permanent dipoles experience both the potential associated with the orientation of the mesogens and that associated with the external electric field.

Going back to Eq. C.14, the potential energy of the molecular dipole of the nonlinear chromophore in a liquid crystalline environment is

$$U(\theta) = U_{\mathbf{s}}(\theta) \; - -\mu E. \tag{C.31}$$

$U_s(\theta)$ contributions to the orientational distribution function influence the degree of polar order that can be obtained in a positive way. To obtain the above expressions for <*cos*>, the value of $U_s(\theta)$ has to be known. Although we do not know it, we can get an idea of what it is by imposing restrictions such as those derived from a very anisotropic environment, for example if we think of an extreme case such as that corresponding to the Ising model. In this scheme, the value of $U_s(\theta)$is equal to zero for $\theta = 0$, π, 2π, etc. and infinite for any other angle. Under these circumstances, the obtained coefficients in the case of low fields are

$$d_{33}\,(\text{Ising}) = 5d_{33}\,(\text{isotropic})\,,$$

$$d_{31}\,(\text{Ising}) = 0\,.$$

For high fields the alignment saturates for both models and $d_{33}\,(\text{Ising}) \rightarrow d_{33}\,(\text{isotopic})\,.$

In practice, we can expect an alignment less rigorous than that of the Ising model. However, the use of liquid crystals appears to anticipate a degree of order higher than that obtained in isotropic materials [10] normally associated with d_{33}/d_{31} ratios clearly greater than three.

Appendix D
NLO Parameter Units

The units we will use for the susceptibility are those of the International System, while the hyperpolarizability and the $\mu\beta$ product are given in cgs units, as is usual in the literature on the subject. Let us see the relationship between the units of the NLO parameters in the two systems.

Second-Order Nonlinear Susceptibility

We will begin by defining the nonlinear polarization in the two unit systems, as expressed in Eq. 1.2:

$$P_{SI}^{NL} = \varepsilon_0 \chi_{SI}^{2(1)} E_{SI}^2 = \chi_{SI}^{2(II)} E_{SI}^2$$
$$P_{cgs}^{NL} = \chi_{cgs}^2 E_{cgs}^2$$

In this thesis the International System with Convention I will be used.

In Table D.1, the units of each parameter in the two systems are presented, together with the conversion factors to go from SI units to the cgs system.

Therefore, by Convention I, the relationship between susceptibility expressed in the two systems will be

$$\chi_{cgs}^{(2)}(esu) = \frac{(3 \times 10^4)}{4\pi} \chi_{SI}^{(2)}(m/V).$$

Throughout this thesis the coefficients are given in pm/V (10^{-12} m/V).

First-Order Hyperpolarizability

In the cgs system the hyperpolarizability β, which relates the induced dipole moment to the squared field, $p = \beta E^2$, is expressed in units of length to the fourth power divided by the unit of electrical potential (cm^4/statvolt).

In the International System, as in the case of the susceptibility, we find two ways to define β:

R. Alicante, *Photoinduced Modifications of the Nonlinear Optical Response in Liquid Crystalline Azopolymers*, Springer Theses, DOI: 10.1007/978-3-642-31756-9,

Table D.1 The units of the parameters PNL, E and χ(2) in the SI and cgs systems of units and their conversion factors

	P^{NL}	E	$\chi^{(2)}$
SI	C/m^2	V/m	m/V
Conversion factor	3 × 10^5	1/(3 × 10^4)	3 × 10^4/4π
cgs	esu[a] (u.e.q.[b]/cm^2)	esu (statvolt/cm)	esu (cm/statvolt)

[a] esu: electrostatic units; their equivalent in the cgs system is given in parentheses
[b] u.e.q.: charge equivalent units

Definition 1: $p = \varepsilon_0 \beta^I E^2$
Definition 2: $p = \beta^{II} E^2$
Therefore $\beta^{II} = \varepsilon_0 \beta^I$
The conversion factors from cgs to SI for β are shown below for both definitions:

$$\beta^I_{SI}(\mathrm{m}^4/\mathrm{V}) = \frac{4\pi}{3\times 10^{10}}\beta_{cgs}(cm^4/statvolt), \text{ so } \frac{m^4}{V} = 2.387\times 10^9 \frac{cm^4}{statvolt},$$

and,

$$\beta^{II}_{SI}(\frac{Cm^3}{V^2}) = \frac{1}{3\times 10^{10}}\varepsilon_0\beta_{cgs}(esu\beta) \text{ so } \frac{Cm^3}{V^2} = 2.7\times 10^{20} esu,$$

where esuβ means the electrostatic units of β, which are cm^4/statvolt as 1 statvolt = 3 × 10^2 V.

As already mentioned, throughout this thesis we will use cgs units, because they are most-often used in the literature. The value of $\mu\beta$ from EFISH measurements often expressed in cgs units (esu). Its units are equivalent to the electrostatic unit of charge (u.e.q.) × centimetre (cm) × electrostatic unit of β (cm^4/statvolt).

To move from $\mu\beta$ in cgs to $\mu\beta$ in SI, one must take into account the relationship between the units of μ in the two systems. In the International System this is expressed in C m, and 1 C m = (3 × 10^9 u.e.q.) × (10^2 cm) = 3 × 10^{11} u.e.q. × cm so that

$$\mu_{SI} = \frac{\mu_{cgs}}{3\times 10^{11}}, \text{ so that } 1\text{ C m} = 3\times 10^{11}\text{u.e.q.} \times \text{cm}.$$

Yet, to convert the $\mu\beta$ values in this thesis (in esu) to SI units (Convention I), we will have

$$(\mu\beta)^I_{SI}\left(\frac{Cm^5}{V}\right) = \frac{1}{(3\times 10^{11})\,(2.387\times 10^9)}\quad(\mu\beta)_{CGS}(esu)$$

In much experimental work μ is given in debyes (D), where 1 D = 10^{-18} u.e.q. × cm

References

1. J. Jerphagnon, S.K. Kurtz, Maker fringes: a detailed comparison of theory and experiment for isotropic and uniaxial crystals. J. Appl. Phys. **41**(4), 1667 (1970)
2. W.N. Herman, L.M. Hayden, Maker fringes revisited: second-harmonic generation from birefringent or absorbing materials. J. Opt. Soc. Am. B **12**(3), 416 (1995)
3. R.L. Sutherland, Handbook of Nonlinear Optics, (Marcel Dekker, New York, 1996)
4. J.L. Oudar, Optical nonlinearities of conjugated molecules. Stilbene derivatives and highly polar aromatic compounds. J. Chem. Phys. **67**(2), 446 (1977)
5. F.J. Rodríguez, *Propiedades ópticas fotoinducidas en polímeros con uidades de azobenceno*. Ph.D. Dissertation, Universidad de Zargoza, 2005
6. M.A. Mortazavi et al., Second-harmonic generation and absorption studies of polymer dye films oriented by corona-onset poling at elevated temperatures. J. Opt. Soc. Am. B **6**(4), 733 (1989)
7. R.H. Page et al., Electrochromic and optical waveguide studies of corona-poled electro-optic polymer films. J. Opt. Soc. Am. B **7**(7), 1239 (1990)
8. V. Rodriguez, F. Lagugné-Labarthet, C. Sourisseau, Orientation distribution functions upon both <P1> <P3> order parameters and upon the four <P1> up to <P4> values: application to an electrically poled nonlinear optical azopolymer film. Appl. Spectrosc. **59**(3), 322 (2005)
9. D.J. Williams (ed.), *Nonlinear Optical Properties of Organic and Polymeric Materials*. ACS Symposium Series, vol. 233, (ACS, Washington, 1983)
10. P.N. Prasad, D.J. Williams, *Introduction to Nonlinear Optical Effects in Molecules and Polymers*. (Wiley, New York, 1991)

Curriculum Vitae

PERSONAL DATA

Name: Dr. Raquel Alicante Santiago
E-mail: raquela@unizar.es

ACADEMIC DEGREES

2011: Ph.D. in physics, "Modificaciones fotoinducidas de la respuesta no lineal en azopolímeros cristal líquido" Condensed Matter Department, University of Zaragoza.
2006: Master's degree in chemistry and polymeric materials, University of the Basque Country.
2002: Graduate degree in chemistry (speciality: macromolecules), University of the Basque Country.

PRIZES

2012: Springer Theses award, recognizing outstanding Ph.D. Research, Springer.
2012: "Outstanding Ph.D award" in Science, Universidad de Zaragoza.

FORMER RESEARCH POSITIONS

2004–2006 Grant funded by "Repsol YPF S.A." and "The University of the Basque Country" for the research "Rheological characterization of polyolefins" in the Rheology Group at the Chemistry Faculty in Donostia-SS, Spain.
2003–2004 Grant funded by "Cintas Adhesivas UBIS S.L." and "The University of the Basque Country" for the research "Synthesis of nanoemulsions of acrylic polymers for P.S.A. adhesives" in the Chemical Engineering Group at the Chemistry Faculty in Donostia-SS, Spain.

R. Alicante, *Photoinduced Modifications of the Nonlinear Optical Response in Liquid Crystalline Azopolymers*, Springer Theses, DOI: 10.1007/978-3-642-31756-9,

2002–2003 Internship at the research laboratory of the company "Cintas Adhesivas UBIS S.L" Hernani, Spain.

CONTRIBUTIONS TO CONFERENCES

- P1: Congress: "International Symposium on Functional pi-Electron Systems", Atlanta (USA); 23–28 May, 2010.
- P2: Congress: "VI Reunión del Grupo Especializado de Física de Estado Sólido", Zaragoza (Spain); 3–5 February, 2010.
- P3: Congress: "XI Reunión del Grupo Especializado en Polímeros", Valladolid (Spain); 20–24 September, 2009.
- P4: Congress: "16th European Symposium of Organic Chemistry", Prague, (Czech Republic); 12–16 July, 2009.
- O1: Congress: "IV Congreso Nacional de Jóvenes investigadores en polímeros", Peñíscola, (Spain); 15–19 June, 2008.
- P5: Congress: "X reunión del Grupo Especializado en Polímeros", Sevilla (Spain); 16–20 September 2007.
- O2: Congress: "European Annual Rheology Conference 2006", Crete (Greece); 27–29 April 2006.
- P6: Congress: "European Annual Rheology Conference 2005", Grenoble (France); 21–23 April 2005.

PUBLICATIONS (articles)

1. "*Matrix Order Influence on the Nonlinear Optical Properties of Dispersed Chromophore-Azopolymer Systems*" *R. Alicante*, R. Cases, B. Villacampa, E. Blasco. *Macromol. Chemical Physics* doi: 10.1002/macp.201100649.
2. "*Understanding Optoelectronic Properties of Cyano-Terminated Oligothiophenes in the Context of Intramolecular Charge Transfer*" Sandra Rodríguez González, Jesús Orduna, *Raquel Alicante*, Belén Villacampa, Kari A. McGee, Joo Pina, J. Seixas de Melo, Kathryn M. Schwaderer, Jared C. Johnson, Brady A. Blackorbay, Jacob J. Hansmeier, Victoria F. Bolton, Tyler J. Helland, Brett A. Edlund, Ted M. Pappenfus, Juan T. López Navarrete, and J. Casado Cordon. *Journal of Physical Chemistry B*, **2011**, 115 (36), 10573–10585.
3. "*Aromatic/Proaromatic Donors in 2-Dicyanomethylenethiazole Merocyanines: From Neutral to Strongly Zwitterionic Nonlinear Optical Chromophores*" R. Andreu, E. Galán, J. Orduna, B. Villacampa, *R. Alicante*, J. T. López Navarrete, J. Casado, and J. Garín. *Chemistry—A European Journal* **2011**, 17, 826–838.
4. "*Benzothiazolium-p-thiazole-dicyanomethanides: new nonlinear optical chromophores*" R. Andreu, E.Galán, J. Garín, J. Orduna, R. Alicante, B. Villacampa. *Tetrahedron Letters* **2010**, 51, 6863–6866.
5. "*Synthesis, electrochemical and theoretical studies of V-shaped donor–acceptor hexaazatriphenylene (HAT) derivatives for second harmonic generation*" R.

Juárez, M. Ramos, J. L. Segura, J. Orduna, B. Villacampa and *R. Alicante*. *The Journal of Organic Chemistry* **2010**, 75, 7542–7549.

6. *"Tailoring Linear and Nonlinear Optical Properties of a Side-Chain Liquid Crystalline Azo-Polymethacrylate"* *R. Alicante*, R. Cases, B. Villacampa, B. L. Oriol. *Macromolecular Chemistry and Physics* **2010**, 211, 2218–2225.
7. *"Isophorone- and pyran-containing NLO-chromophores: a comparative study"* R. Andreu, S. Franco, E. Galán, J. Garin, N. M. de Baroja, C. Momblona, J. Orduna, *R. Alicante*, B. Villacampa. *Tetrahedron Letters* **2010**, 51, 3662–3665.
8. *"Bent-core liquid crystals in a route to efficient organic nonlinear optical materials"* I.C Pintre, J. L. Serrano, M. B. Ros, J. Martinez-Perdiguero, I. Alonso, J. Ortega, C. L. Folcia, J. Etxebarria, *R. Alicante*, B. Villacampa. *Journal of Materials Chemistry* **2010**, 20, 2965–2971.
9. *"Linear and V-Shaped Nonlinear Optical Chromophores with Multiple 4H-Pyran-4-ylidene Moieties"* R. Andreu, E. Galan, J. Garin, V. Herrero, E. Lacarra, J. Orduna, *R. Alicante*, B. Villacampa. *Journal of Organic Chemistry* **2010**, 75 1684–1692.
10. *"Synthesis and Nonlinear Optical Properties of Side Chain Liquid Crystalline Polymers Containing Azobenzene Push–Pull Chromophores"* *R. Alicante*, R. Cases, P. Forcen, L. Oriol, B. Villacampa. *Journal of Polymer Science Part A* **2010**, 48, 232–242.
11. *"4H-Pyran-4-ylidenes: Strong Proaromatic Donors for Organic Nonlinear Optical Chromophores"* R. Andreu, L. Carrasquer, S. Franco, J. Garin, J. Orduna, N. M. de Baroja, *R. Alicante*, B. Villacampa, M. Allain. *Journal of Organic Chemistry* **2009**, 74, 6647–6657.
12. *"Synthesis and nonlinear optical properties of chromophores for photorefractive polymer materials"* E. Font-Sanchis, F. J. Cespedes-Guirao, A. Sastre-Santos, B. Villacampa, J. Orduna, *R. Alicante*, F. Fernandez-Lazaro, *Tetrahedron* **2009**, 65, 513–4520.
13. *"New one- and two-dimensional 4H-pyranylidene NLO-phores"* R. Andreu, L. Carrasquer, J. Garin, M. J. Modrego, J. Orduna, *R. Alicante*, B. Villacampa, M. Allain. *Tetrahedorn Letters* **2009**, 50, 2920–2924.
14. *"Decreased Optical Nonlinearities upon CF3 Substitution on Tricyanofuran Acceptors"* R. Andreu, M. A. Cerdan, S. Franco, J. Garin, A. B. Marco, J. Orduna, D. Palomas, B. Villacampa, *R. Alicante*, M. Allain. *Organic Letters* **2008**, 10, 4963–4966.
15. *"Iminium salts of omega-dithiafulvenylpolyenals: An easy entry to the corresponding aldehydes and doubly proaromatic nonlinear optic-phores"* S. Alias, R. Andreu, M. J. Blesa, M. A. Cerdan, S. Franco, J. Garin, C. López, J. Orduna, J. Sanz, *R. Alicante*, B. Villacampa, M. Allain. *Journal of Organic Chemistry* **2008**, 73, 5890–5898.
16. *"Electronic, optical, and vibrational properties of bridged dithienylethylene-based NLO chromophores"* M. C. R. Delgado, J. Casado, V. Hernandez, J. T. L. Navarrete, J. Orduna, B. Villacampa, *R. Alicante*, J. M. Raimundo,

P. Blanchard, J. Roncali. *Journal of Physical Chemistry C* **2008**, 112, 3109–3120.

17. *"NLO properties of dithienothiophene-based chromophores: A comparison study between the Donor/Donor and Donor/Acceptor substitution patterns*"-art. no. 66530X. M. C.R. Delgado, J. Casado, V. Hernandez, J.T.L. Navarrete, J. Orduna, B. Villacampa, *R. Alicante*, J. M. Raimundo, P. Blanchard, J. Roncali, Conference on Linear and Nonlinear Optics of Organic Materials VII, Aug. 28–30, **2007**, San Diego, CA, .

18. *"New results on the correlation molecular architecture–melt elasticity–blowing process–film properties for conventional and metallocene catalysed polyethylenes*" A. Santamaría, *R. Alicante*, R. Pérez, V. Leal, P. Lafuente. *Macromolecular Materials and Engineering* **2006**, 291, 670–676.

19. "*Binary and ternary mLLDPE blends: Tailoring their properties using complex viscosity measurements*" M. A. Flores, *R. Alicante*, R. Pérez, E. Rojo, M. Fernández, A. Santamaría. *Macromolecular Materials and Engineering* **2005**, 290, 704–709.

FSC
www.fsc.org
MIX
Papier aus verantwortungsvollen Quellen
Paper from responsible sources
FSC® C105338

If you have any concerns about our products,
you can contact us on
ProductSafety@springernature.com

In case Publisher is established outside the EU,
the EU authorized representative is:
Springer Nature Customer Service Center GmbH
Europaplatz 3, 69115 Heidelberg, Germany

Printed by Libri Plureos GmbH
in Hamburg, Germany